# Secularism and Its Opponents from Augustine to Solzhenitsyn

# Secularism and Its Opponents from Augustine to Solzhenitsyn

Emmet Kennedy

palgrave
macmillan

SECULARISM AND ITS OPPONENTS FROM AUGUSTINE TO SOLZHENITSYN
Copyright © Emmet Kennedy, 2006.
All rights reserved. No part of this book may be used or reproduced in any manner whatsoever without written permission except in the case of brief quotations embodied in critical articles or reviews.

First published in 2006 by
PALGRAVE MACMILLAN™
175 Fifth Avenue, New York, N.Y. 10010 and
Houndmills, Basingstoke, Hampshire, England RG21 6XS.
Companies and representatives throughout the world.

PALGRAVE MACMILLAN is the global academic imprint of the Palgrave Macmillan division of St. Martin's Press, LLC and of Palgrave Macmillan Ltd. Macmillan® is a registered trademark in the United States, United Kingdom and other countries. Palgrave is a registered trademark in the European Union and other countries.

ISBN-10: 1-4039-7615-5     hardcover
ISBN-13: 978-1-4039-7615-4  hardcover

Library of Congress Cataloging-in-Publication Data

Kennedy, Emmet.
  Secularism and its opponents from Augustine to Solzhenitsyn/Emmet Kennedy.
    p. cm.
  Includes bibliographical references and index.
  ISBN 1-4039-7615-5 (alk. paper)
    1. Secularism–History. I. Title

BL2727.8.K46 2006
211'.609–dc22                                                           2006043285

A catalogue record for this book is available from the British Library.

Design by Macmillan India Ltd.

First edition: 2006

10 9 8 7 6 5 4 3 2 1

Printed in the United States of America.

To
Dan and Rob
Who were willing and able to help

Those two cities are interwoven and intermixed in this era, and await separation at the last judgement.

Augustine, *The City of God*

The line dividing good and evil cuts through the heart of every human being . . . This line keeps changing place; . . . it is after all because of the way things worked out that they were the executioners and we weren't.

Aleksandr Solzhenitsyn, *The Gulag Archipelago*

# Contents

| | | |
|---|---|---|
| Acknowledgments | | ix |
| One | Introduction | 1 |
| Introduction Part I: Augustine and Aquinas | | 11 |
| Two | Saint Augustine: Christianity and the Roman World | 13 |
| Three | Thomas Aquinas's Christian Secularism | 23 |
| Introduction Part II: Renaissance and Reformation | | 39 |
| Four | Dante and Lay Secularism of the High Middle Ages | 41 |
| Five | Machiavelli, Religion, and Politics | 57 |
| Six | Luther's Centrifugal Reformation | 73 |
| Introduction Part III: Autonomy in the Enlightenment | | 91 |
| Seven | Locke: Toleration, Infallibility, and the Secular State | 93 |
| Eight | Rousseau: The Secular Hermit | 113 |
| Nine | Immanuel Kant's Ambiguous Secularism | 131 |
| Introduction Part IV: The Dialectic Upward and Downward | | 145 |
| Ten | G. W. F. Hegel: Secular Philosophy Comprehending Theology | 147 |
| Eleven | Marx: "The Christian State," "The Jewish Question," and "Species-Being" | 163 |
| Introduction Part V: Reaction against Secularism | | 181 |
| Twelve | Dostoyevsky and European Secularism | 183 |
| Thirteen | Solzhenitsyn, Communism, and the West | 203 |
| Fourteen | Conclusion | 225 |
| Notes | | 231 |
| Index | | 269 |

# Acknowledgments

Secularism is a touchy topic. I want to thank all those individuals who came from so many philosophical and religious persuasions to help me with what must have seemed a risky subject. My wife, Janie, is a wonderful mistress of the English language and scrutinized many a troublesome passage. My brother, Malcolm, encouraged me after reading half the manuscript. My sons, Daniel and Rob, showed a faith in the project and performed numerous chores that kept me going. To them the book is dedicated. Jonathan Chaves, friend, colleague, and sinologist, who knows Western thought better than most Europeanists, if not most Europeans, read the entire manuscript, offered numerous suggestions, and gave me his unflagging support. Pierre and Beatrix Audigier lent me their apartment in the summer of 1996, and it was there that I began my studies. My neighbor Bill Coyle provided valuable computer expertise. James Truscott, John Headley, and Dewey Wallace gave the chapters on Dante, Luther, and Locke close readings, respectively. John Marshall's Folger Library seminar on Locke in 2005 helped me refine my views on that pivotal philosopher. Harvey Chisick ferreted out some errors in the Rousseau chapter and discussed with me the question of whether there are any alternatives to secularism. Joshua Konecni lent a keen editorial eye to the same chapter. The German and Russian education of my colleague Peter Rollberg made his critique of Chapters 12 and 13 authoritative. Likewise, another Slavic professor, Alex is Klimoff of Vassar College, gave me valuable information on both Dostoyevsky and Solzhenitsyn on several occasions. John Ziolkowski assisted me with his Latin and computer expertise. Daniel J. Mahoney, whose book on Solzhenitsyn was an inspiration, read and appreciated the entire manuscript with exceptional enthusiasm. Louis Dupre generously commented on Chapters 7, 8, and 10–12. Others provided valuable hints, references, and suggestions: they are Judy Plotz, Lynn Hunt Dave Gallagher, Harvey Mansfield, the late R. R. Palmer, Orest Ranum, Martin Staum, William Stetson, Ronald Gillis, Lawrence Kutz, Carl Schmitt, and J. G. A. Pocock. Howard Sachar gave me constant encouragement and editorial assistance.

Much credit goes to Trish Bozzell, one of Washington's top editors, who combed carefully through the manuscript for more precise and more economical ways of expressing my ideas. A number of students and former students served as excellent proofreaders and research assistants. They are

Jason Kirwan on Solzhenitsyn, Paul Du Quenoy on Dostoyevsky, and Jeffrey Burson on the notes. The younger and more computer-literate generation worked through several drafts of proofreading and correcting. Among them were Liz Zentos, Catherine Bosley, Barbara Fox, Varad Mehta, Michael Weeks, Artemy Kalinovsky, Brendan Coyle, Bestami Bilgic, and Greg Desforge. The initial research for the book was made possible by a grant from the George Washington University Facilitating Fund and later grants-in-aid from the history department and the Elliott School of International Affairs at George Washington University. The libraries whose resources provided the rich materials for the book were the Washington Consortium of Libraries, the Catholic University's Philosophy, Theology, and Humanities libraries at the Mullen Library, as well as George Washington's Gelman Library and its Reference, Circulation, and Inter-Library Loan divisions—all of whose librarians were constantly helpful. A great deal of the work was done at the largest library in the world with the help of its expert reference staff—the Library of Congress.

I am indebted to the Göteborg (Sweden) Enlightenment colloquium, which published a paper of mine on the Enlightenment origins of postmodernism, and to the forum of graduate students at Justus Liebig University (arranged by Rolf Reichardt and Gerhard Löttes), where I gave an earlier version of the Marx paper. Blackwell Publishers kindly granted permission for publication here of that paper, originally scheduled to appear in the *Journal of the Historical Society*. The Western Society of French History hosted and published in its *Proceedings* vol. 28 (2000), a paper on "Rousseau's Religiosity and Secularism." At the Colloquium on the Ideologues in Tübingen, Germany, I gave a paper on the "Secularism of Destutt de Tracy," which is to appear on its website. None of the above is responsible for the interpretations that follow.

# Chapter One
# Introduction

Throughout most of human history, religion has been intimately bound up with government. Separation of church and state is a modern exception. In the Greco-Roman world, the priest was an official whose task was to win the favor of the gods, for they alone could guarantee the city's survival. Liberty was the right to participate in politics—a freedom of the agora and the forum, not of thought or religion. The coming of Christ dramatically changed the ancient relationship between religion and state, priest and official. It was he who announced, "My kingdom is not of this world," and said, "Render unto Caesar what is Caesar's and to God what is God's." His disciples were to bid all nations to follow him—ultimately, into another world. Initially they were persecuted by the Jews, who saw Christ as an impostor, and then by the Romans, who regarded his followers as seditious, since they worshipped a king other than Caesar. Like the Jews, Christians were regarded as the "odium of the human race," because, unlike the rest of men, they refused Rome's idols. And indeed, when Rome fell in A.D. 410, Christians were held responsible for having deserted the Roman tutelary gods. This was the occasion for that notable Roman orator, and North African bishop, Augustine of Hippo, to compose a lengthy defense of Christianity, which he both defended and convicted inadvertently. He insisted that Christians bear their share of the civic burden of the empire. But he identified the latter as "a terrestrial city," which was distinct from and more transitory than the city of God, peopled by true followers of Christ. Like Christians before him, Augustine tended to identify the *saeculum* (the Latin word for world, century, or age) and *saeculare* (the adjective derived from it) with the terrestrial city.[1]

Today the derivatives of the word *saeculum* connote absence of religious feeling, a worldly rather than an otherworldly approach to life. In the nineteenth century, secularism denoted the exclusion, or at least the reduction, of the role of religion in public life, in politics, in education, and in all branches of learning. It is the cultural counterpart of the separation of church and state. It may imply a disbelief or skepticism in the supernatural, which is considered nonexistent or too tendentious to warrant outward recognition. Or it may imply a disagreement about particulars (such as doctrines or forms of worship), resulting in a multiplication of sects, negating the feasibility or justifiability of establishing any one of them. Those considerations have guided the United States and France directly, and other Western countries indirectly, in their determination to separate religion from public life, and most recently, in the 2004 decision of the EEC, to refuse John Paul II's request for an acknowledgment of its Christian roots in the European constitution. The advantages are patent. No one religion can pretend to speak for the rest.

But there are disadvantages as well. This book will show the cost of secularism, rather than its benefits (which are generally conceded)—the side effects of an otherwise beneficial institution. As a history it does not provide ready-made answers. It aims, rather, at showing why secularism is something of a problem, rather than a solution. The separation of church and state is certainly an issue. But I am more concerned with the ideas and culture of secularism than with constitutional arrangements. The ideas of religious nonconformists and secular dissenters in regimes of opposite tendencies are crucial. Thoughts are here shown to be intertwined with lives. Many thinkers in the following pages—Dante, Machiavelli, Locke, Rousseau, Marx, Dostoyevsky, and Solzhenitsyn—were exiled or imprisoned for their beliefs.

These thinkers fall into three categories: Augustine and Aquinas set the terms of the debate between Christian and non-Christian culture, but were not secularists. (They both, however, use the words *saeculare* and *saeculum* frequently.) These two doctors of the church spoke to the problem of the relationship of reason to faith (which also concerned Rousseau, Kant, and Hegel), as well as to the church and state issue. Other thinkers (Dante, Machiavelli, Locke, Rousseau, and Marx) criticized papal power and the role of religion in politics, which is certainly a development of the secular doctrine. Secularism was, of course, a strong component of French revolutionary ideology, which spread over Europe after 1789. That event marked Hegel's and Marx's ideas indelibly. The last thinkers we consider—Dostoyevsky and Solzhenitsyn—reaped the results of centuries of this antireligious criticism and, in their works, linked secularism to the scourge of communism by prophecy and experience respectively. Both were sharp critics of secularism. They vindicated the rights of religion, which was no longer the persecutor, but the persecuted.

Thus, rather than studying secularism as a disembodied idea, I have traced it through recognizable individuals who, more than most of their contemporaries, spoke for their age. The phenomenon, as two colleagues William Burrows and R. R. Palmer have pointed out, "is like the air we breathe," something that "can be found everywhere" and hence in many other thinkers. I could have chosen, for instance, less well-known figures such as Orosius, Albert the Great, Guicciardini, and Soloviev, or giants such as Hobbes, Voltaire, Nietzsche, and Freud. My selection is somewhat arbitrary, which is to acknowledge the ubiquity of secularism. Let others follow who can write other histories.

The intellectual and physical contexts in which all the works I examine below were written are crucial. Augustine might not have thought of the dichotomous two cities if he had not lived in Rome and Hippo. St. Thomas could not have written the two *summae* without the recently translated works of Aristotle. Machiavelli's political theory is incomprehensible if viewed apart from the violent political practices of the Renaissance city-state, the apocalyptic Savonarola, and the papacy. Luther's attitude toward the papacy might have been more conciliatory had he not been excommunicated. Locke's views on tolerance were the result of long cogitations on civil war and sectarianism, which called into question the very existence of "orthodoxy." Kant must be judged against the influence of David Hume's skepticism and empiricism. Marx appears much more reasonable when juxtaposed against the Industrial Revolution in Britain. The nihilism that Dostoyevsky rebelled against and the Stalinism that Solzhenitsyn ultimately triumphed over were hurdles for titans as well as for true believers.

The Middle Ages (400–1300) witnessed a victory of the sacred over the secular. But ever since then, the sacred has been in erratic retreat, often led by the intellectuals on whom this book focuses. The church fathers relied on faith over reason, the sacred over the secular. Thus Augustine's *fidens quaerens intellectum* (faith seeking understanding) meant that one must believe in order to know, which, in the eleventh century, was epitomized by Anselm of Canterbury as *credo ut intelligam* (I believe that I may understand). But it is in the High Middle Ages (1100–1300) that elements of modern rationalism begin to stir.

Thomas Aquinas created two parallel systematic treatises, one of reason (*Summa contra Gentiles*) and the other of faith (*Summa theologica*). While the division was not watertight, in the first he relied mostly on classical, pre-Christian sources, and in the second mostly on Christian theological ones. In other words, he recognized a distinction between reason and revelation, and this was momentous. He differs from Descartes, who would also make such distinctions, in that he considered theology based on revelation feasible and superior to reason, whereas Descartes enclosed religion in a parenthesis he did not open even though he accepted its contents. This respectful suspension by Descartes is secularism in that it segregates religion from philosophy. But Aquinas was the first to make the distinction, the difference being that for him theology was knowledge, a superior knowledge to boot, albeit a knowledge based on an other principle, namely revelation.

The centuries after Aquinas—the Renaissance centuries—show a further gulf separating reason from religion, represented by an ever more powerful papacy, and the emerging state, represented by the Western medieval monarchies, the Renaissance city-state, and the ailing Holy Roman Empire. The focus thus changes from the Thomistic reason and revelation (which shaped Dante's *Divine Comedy*) to the relation of church and state. Both Dante's and Machiavelli's views are examples of pre-Reformation challenges to papal temporal power.

The Protestant Reformation is the key backdrop to modern secularism, for Luther tied personal salvation to scripture (*sola scriptura*) and faith (*sola fides*) rather than to church authority or good works. But it did not produce a consensus. Not only did Luther break with the Catholic Church, but his followers and imitators broke with him and with each other. The result was the splintering of Christendom and the establishment of a religiously plural Europe and America, the source of our present-day diversity and secularism, which accommodates many individual consciences, but it does so with a continuing splintering that necessarily makes any kind of religious unity and community seem remote. Such religious divisions ultimately lead to the repression of outward expression of religious sentiments in contemporary society, lest they elicit discord. This repression is one of the key manifestations of secularism. Voltaire noted, in his *Letters on England,* that Muslims, Christians, and Jews get along fine at the stock exchange, where religious affiliations are irrelevant. There, one ratchets down to a lower common denominator, where men can presumably live without conflict, but, we would add, also without much meaning.

The Enlightenment can be characterized by several things—reason, reform, critical spirit, hostility toward revealed religion. We look at it from the perspective of autonomy and individualism. It has been convincingly argued and documented that the skepticism of the sixteenth and seventeenth centuries, most notably that of Michel Montaigne (1518–1585), was the result of Catholic-Protestant controversy and war. John Locke concluded, after viewing with dismay the sectarianism of the English civil wars of the 1640s, that "everyone is orthodox to himself," and that governments should therefore not meddle in religion. Religion would thus devolve upon the individual. There could be no public faith—neither of Christendom nor of a nation—because no church, he felt, could claim to be "the true church," without a rival nullifying its claim. Self-reliance, or autonomy, then, became a watchword of the Enlightenment.

Rousseau advocated, in *The Social Contract*, a Roman republican religion for the state, relegating inner religiosity (that of the Gospel—a "religion of man") to the individual. The former, though politically useful, could satisfy man's religious yearnings no more than could a parade. The autonomous individual must seek God in another, more spiritual domain. To achieve this, Rousseau, like many of his generation, eliminated revelation and miracles—any divine constraint in the world was impossible. Man was really autonomous. The Gospel, for him, as for Jefferson and other deists, was an unmatched compendium of moral truths, not a revelation of God's mysteries or commands. Indeed, Rousseau's moral and political theory is based on obedience to oneself.

The medieval Spanish Arab Averroes had advanced the notion of a "double truth"—one for philosophy, one for theology. These truths contradicted each other, yet each was true in its proper domain. Aquinas, on the contrary, insisted on the unity of truth, since God, the author of truth, cannot contradict himself. Averroes's doctrine of the double truth can be found in the Enlightenment, particularly in Voltaire, who advocated belief in God for his servants but felt free to doubt his existence himself. Kant's *Critique of Pure Reason* (1781) and *Critique of Practical Reason* (1788) follow an Averroistic train of thought. In the first *Critique*, Kant argues that God's existence and that of an afterlife and freedom of the will cannot be indisputably proven. In the second *Critique*, he argues that those same beliefs are necessary to live a moral life. It is quite likely, however, that any readers who pride themselves in intellectual integrity will eschew such "practicality" and conclude that there is very little reason to believe in God. Like Hume, Kant is a crucial source of modern atheism and agnosticism. He is also the inspiration of nineteenth- and twentieth-century European Protestantism, which shifts more and more of the onus of Christian credibility to faith and inwardness as opposed to what is regarded as Catholic dogmatism and externalism.

Kant's critiques themselves are more secular than atheistic. The double truth lets him refute theism and then endorse it on different grounds—a subtle, secular solution. The impact of the refutations of God, immortality, and freedom in the *Critique of Pure Reason* has been historically more influential than their disclaimers and qualifications in the *Critique of*

*Practical Reason.* The split between the two sundered the unity of inherited philosophical thought.

In the section titled "The Dialectic Upward and Downward," I show how Hegel converts the Kantian "antinomies," or contradictions, into a dialectic that restores the importance of externality and objectification. Hegel, for his part, stresses that the idea never fulfills itself until it comes into conflict with its antithesis and rises up to a new level. Thus the Enlightenment and the French Revolution laid too much emphasis on individual reason and right, which led to anarchy. They can only be made viable when they are solidified in a viable state. For Hegel, the state takes the place of the church, which, in his *Philosophy of Right* (1821), is morally and institutionally subsumed by the state. This exceeds the Lutheran subordination of the church to the state. What is noteworthy in Hegel and his follower, Marx, is that all confidence in redeeming man has shifted from the church to the state and that the nature of this redemption itself becomes wholly secular. It mirrors the numerous "secularizations" of church lands by the state throughout revolutionary and Napoleonic Europe.

Hegel believes that man's fulfillment is reached not only through transcendence of contradictions and objectification in institutions, but also through a growth of self-consciousness at each step. The result in Hegel is to bring God's existence to reside exclusively in human consciousness, which, as "Absolute Spirit," then takes his place. When Hegel writes of Christ's Incarnation, one is left wondering whether God became man or man became God. Indeed one of his followers wished he would speak more clearly of God as a really existent thing and not just as a thought in human consciousness. The problem has been described as "philosophy comprehending theology," or reason understanding God—the possibility of which Christian philosophers, medieval as well as modern, have been loathe to admit. (Aquinas had thought one could demonstrate God's existence but not understand his essence.)

Marx disliked Hegel's spiritualization of reality and steered toward pure materialism during his studentship in Berlin in the 1830s. The elements of his dialectical materialism (an inversion of Hegel's spiritual dialectic) are well known. Shifting his emphasis from spirit to human society, he sees the real cause of human alienation and misery to spring from the conflict of social classes, the dominant ones exploiting those underneath. Private property, of course, was one root of the problem, but religion was a less noticed one. Religion segregates humanity into denominations, instead of uniting it on the basis of "species being," or what one holds in common with one's species. Private property and sectarian religion both alienate the individual from humanity. For Marx the solution is common property. In his 1843 article "On the Jewish Question," he argues for the abolishment of religion rather than for the establishment of a common one. This measure was incorporated, of course, by communist governments in the twentieth century. The Bolshevik Revolution of 1917 marks the apogee of secularism, because human association was then based not according to what is highest—beliefs and religion—but according to what is lowest—physical existence.

Other nineteenth-century populisms and pantheisms such as Auguste Comte's "Great Being—Humanity," Victor Hugo's *Misérables,* George Sand's *waifs,* Eugene Sue's *urchins,* and Michelet's *people* glorify humble folk at the expense of their bourgeois antagonists, who were well punished by Honoré Daumier's brush. The hierarchy of human perfection had been turned upside down. Ludwig Feuerbach went so far as to argue that God was really only man's image of himself objectified. God was now man and man was God. Many nineteenth-century thinkers had been captivated by a secular immanence, unable to conceive of an order above their own.

Russia took the "new faith" most seriously just when France experienced its last revolution (the Paris Commune of 1871). Before Nietzsche would reach the opinion that "God is dead," Dostoyevsky was dramatizing some of its consequences in a novel called *Demons* or *The Possessed* (1871). In the book a provincial town observes atheism coalesce with professional revolutionism. Hegelianism and the latest intellectual fashions from the West are trumped by an underground revolutionary network that finally torches part of the town and murders some of its inhabitants. The novel, which captures midcentury Russian terrorism, populism, and nihilism, can be seen as a prophecy of totalitarianism. Dostoyevsky's Slavophile novels are among the first and most powerful critiques of Western secularism.

The greatest Russian opponent of communist secularism is Aleksandr Solzhenitsyn (1918–). His writings all reflect his incarceration in the gulag (1945–1953)—the immense Soviet labor camp that stretched across Siberia. He believes that the essence of communism is atheism—all the rest is window dressing. However, we in the United States were fixated on Russia's nuclear arsenal and foreign policy. Solzhenitsyn wove together a life of resistance after Stalin's death. He then set about exposing the horrendous penal system with a novella, *One Day in the Life of Ivan Denisovich* (1962), and one immense trilogy, *The Gulag Archipelago* (1973–1976). Smuggling the latter out of the Soviet Union involved taking tremendous risks, including the creation of a clandestine network of collaborators. Exiled in 1974 after its publication in the West, he eventually took up residence in

Vermont until he returned to Russia in 1994. He spoke frequently all over Europe and the United States, most notably at Harvard in 1978, where his analysis of the ideological and spiritual malady of the United States led him to issue a warning: 1917 was repeatable. His Harvard speech was shocking yet perspicacious, and he failed to perceive the extent of his own influence on the collapse of communism, wrapped up as he was in "warning the west" of its menace and the inadequacy of détente. Yet he realized that totalitarianism can take many shapes. He recognized the threat of American advertising (to sanity?), the uncritical subscription of academics to post-Renaissance humanism, and the skewed disinclination to recognize worthwhile journalists and academics from nonprestigious institutions. But he failed to foresee the impact of postmodernism or the belief that there is no truth, and that if there is no truth one should logically stop teaching and writing as if there were. Academia was striking at the heart of traditional values and threatened to create an intelligentsia that C. S. Lewis warned would be cynical and unfit to teach. Perhaps we are faced with two potential totalitarianisms, fundamentalism and secularism, neither of which wants to tolerate the other. The controversy over secularism still rages, in fact rages more strongly. There are those who feel, like Marx, that our political or legal society is secular—adamantly so—and that our civil society is no less adamantly religious—fundamentally so. The victory of the misnamed "red states" in the 2004 U.S. elections over the "blue states" was, in part, a victory of religious and moral values over secularism.

The terrorist attacks of September 11, 2001, provoked for their part an evanescent recourse to things spiritual in order to make sense of mindless death and sudden insecurity—perhaps the wellspring of religion. The net effect has been a retreat of secular discourse and an advance of religiosity in the public sphere. This is disturbing to secularists, but it is probably the more authentic register of how the country thinks.

I do not cover America or the Islamic world in this book, but a few words can be said. Recent decades have produced a crisis in the relative balance of secularism and fundamentalism in the Islamic world. Originally this conflict was largely the doing of the West, which exported secularism to the colonized world along with railroads, electricity, and high-rises. Militant Islamic fundamentalism is partly a reaction to the "Death of God" in the West, which was perceived as a threat in the Maghreb, Turkey, and Iran. As long as the West is perceived as "godless," terrorism will find an excuse to continue.

There remains Europe, the focus of my study. Low church attendance, prohibition of signs of religion in French schools, a refusal to intervene significantly and early in genocidal outbursts on their own and nearby continents (e.g., Bosnia and Sudan), the almost exclusive economic preoccupations of the European Union, the crisis of marriage or lack thereof, and in general the phobic exclusion of religion from the discourse of today's "bien-pensants" are all signs of a secularization of the European mind since the Second World War.

I have found no general history of secularism that reaches back to the origins of Western civilization and forward to our time. Most histories treat the phenomenon as exclusively contemporary. It *is* a contemporary phenomenon, responsible in part for some of the great tragedies of our age. But its roots sink deep into the past.

# Part I

## Introduction:
## Augustine and Aquinas

Aristotle wrote in *The Politics,* "He who is without a polis, by reason of his own nature and not of some accident, is either a poor sort of being, or a being higher than man." The polis in classical Greece was designed to secure man against all the vicissitudes of nature and human life. Christ promised such security only in a higher life.

The juxtaposition of Jesus with Aristotle points to the foundational question of this book: How could one be a Christian and still be a citizen of this world? Are secularity and Christianity compatible? When Rome was sacked in A.D. 410, Christians were blamed because they had abandoned the pagan shrines that had protected Rome for centuries. In withdrawing from its temples, Christians had symbolically withdrawn from Rome. Or had they? Augustine, to Christianity a convert and a bishop, took up the pen in the decade following the sack and wrote a European classic, *The City of God.* In this work he argued that a Christian *could* be a citizen of two cities—the city of God and the terrestrial city. The *saeculum* was the world wherein men forged their ultimate destinies in one of these two cities.

Augustine did not give up hope for civilization simply because Rome had been sacked. He had more confidence in the salvation of barbarians than did Romans in their civilization. And it was, of course, the more perspicacious outlook: the Roman Empire *did* disintegrate in the West; the state was led by "gangs of criminals on a large scale"; and a dark age of economic primitiveness and illiteracy followed, but the church ultimately converted and civilized the barbarians.

Quite different was the thirteenth century, which saw the revival of towns and trade, the growth of population and prosperity, the birth of universities and development of medieval scholasticism, the reconquest of Muslim Spain, and the Crusades to the Near East. These were modernity's birth pangs. It was during these years that Aristotle, recently rediscovered, posed serious problems to Christian philosophers. But Thomas Aquinas assimilated this pre-Christian thought courageously and optimistically in his two *summae*. Overall, his work instilled confidence in human nature, in reason, and in the terrestrial city—the feudal monarchies and the Italian republics.

CHAPTER TWO

SAINT AUGUSTINE: CHRISTIANITY AND THE
ROMAN WORLD

i

By the third century A.D., the ancient city could no longer give its people what they had come to expect. A new element had been introduced: Christianity, a worldview that directed men to think inwardly of immortality and outwardly of eternity. Augustine of Hippo (A.D. 354–430), who has been called "the first modern spirit, one of the founders of modern Western consciousness," was the great apologist of the new religion. He was born in Thagaste in modern Algeria, which was then part of the Roman Empire. His father, Patricius, was a poor squireen, and his mother, Monica, a devout Christian who spent twenty years praying for her son's conversion. Thanks to a neighbor's benevolence, Augustine received a good Roman education in Carthage.[1]

In A.D. 383 Augustine traveled to Rome and Milan, where he gained recognition for his excellence as a rhetorician in patrician society. In these years of high living, he dabbled in Manichaeanism but came into contact with two Christians: St. Jerome, the compiler of the Latin Vulgate version of the Bible, and St. Ambrose, then bishop of Milan. Both had great influence on his conversion. Augustine kept a mistress, but composed a prayer: "Give me chastity, . . . but not yet." He continued to waver, hesitant to take the final trusting step. Then, one sunny day in a Milanese garden, he was sitting with the Bible on his lap when he heard a child pronounce the words *Tolle lege, tolle lege* (Take and read, take and read). Augustine was mystified: he knew neither the voice nor from where it came. But he took up the Bible and opened randomly to Romans 13:13: "Lay aside the works of darkness, and put on the armor of light . . . not in revelry and drunkenness, not in debauchery and wantonness, not in strife and jealousy." The verses, he thought, spoke to him. He was baptized; wrote his *Confessions,* a landmark of spiritual introspection; and in A.D. 391 returned to North Africa, where in due course he became the bishop of Hippo. There he wrote innumerable tracts unflaggingly defending Christianity against Manichaeans, Pelagians, and Donatists.[2]

Secularism concerns man's attitude and relation to the world. Augustine combated several Christian heresies that addressed this issue, and in doing so, he defined the first truly nuanced Christian view on the matter. Manichaeanism had pre-Christian roots, but by Augustine's time it was regarded as a heresy. Its basic tenet was that there are two coeternal, equally powerful principles in the world—the good, or spirit, and evil, or matter (i.e. the body). This tenet led ineluctably to disparaging sex and marriage as being contrary to human perfection. Thus the Manichaean elect were celibate. Manichaeanism denigrated free will and human responsibility; virtue and vice, it held, were largely beyond human control. The attraction of Manichaeanism for Augustine was its stress on irresponsibility. The converted Augustine differed from his erstwhile Manichaean mentors: the flesh itself was good, because God created it, and because Jesus Christ took it on as the incarnate God. Augustine repeated several times in the thirteenth book of the *Confessions* that God saw that creation was good. The origin of evil was not carnal, Augustine stated, but intellectual pride and the seduction by the devil, who promised Adam and Eve that they would be "like gods" if they ate of the tree of life. Ever since, man has inclined toward evil; only Christ's redemption can save him from that original sin. Henceforth, nothing will satisfy Augustine short of the love of God.[3]

Augustine refuted all aspects of Manichaeanism in many works (e.g., *De moribus ecclesiae catholicae et de moribus Manichaeorum*). The Christian account of creation, for instance, led him to believe that Adam and Eve could have had legitimate intercourse before the fall, when their bodies were submissive to their wills. Even subsequently, nothing was inherently evil, certainly not the *saeculum* (since it was created by God); it was only tainted by evil through man's fall.[4]

As a bishop, Augustine also confronted Pelagianism, a heresy that exaggerated man's free will by positing his immunity to the effects of original sin and by stressing his ability to merit heaven on his own. It was almost a contradiction of Manichaeanism. Grace, or the divine bending of the will toward God, was not a tenet of the Pelagians. For Augustine it was the indispensable means of salvation as it would be for his theological offspring throughout time. Furthermore Pelagianism would make Christ's death on the cross superfluous rather than the ultimate source of grace. The

Pelagians, he wrote, "glorify themselves, [but] . . . a man cannot live right, unless he is divinely assisted by the eternal light."[5]

On still another flank were the Donatists. During Diocletian's major persecutions of Christians (A.D. 303–313), many of them, including priests, apostatized to save their lives and, after the persecutions, sought reconciliation with the church. But the rigorist Donatists refused to grant them readmission—not even acknowledging their penitence—and denied the legitimacy of the sacraments administered by their clergy.[6]

These heresies of the early church served to hone Augustine's dualism: the city of God versus the terrestrial city—one inhabited by those in love with God, the other by those in love with themselves. The Pelagians belonged to the terrestrial city, for they believed that human merit was sufficient for salvation. If they reached the city of God, it was by their own efforts. The Donatists believed they were already justified inhabitants in the city of God, in contrast to the lapsed Catholics.

Put another way, the Pelagians were the ancestors of the Renaissance individual and the twentieth-century existentialist, both of whom claimed to be what they made themselves. The Donatists, on the other hand, were the ancestors of the seventeenth-century Jansenists, who believed in the divine election of only a few.

The Manichaeans saw the worldly, sensualist living for pleasure and fame, which they argued are the root of evil. But Augustine did not believe that carnal appetite was the principle of all evil. Rather, he located it in the eye of the soul turning inward upon itself in self-admiration. Nothing will satisfy it, however, except that for which it was made: "The heart is restless, until it rests in thee." This is Augustine's main idea—a psychological confession of the existence of God.[7]

Love of self is part of a more general love of the world. Augustine's outlook on the world (*saeculum*) reflects the Gospel of St. John: One must be in the world, but not of the world. While man inhabits the ancient city, sharing every legitimate worldly pursuit, he in truth belongs to a world beyond. Two centuries before Augustine, the Christian use of the word *saeculum* was pejorative, designating a hostile, corrupt environment. So too was Augustine's, whose attitude toward the *saeculum* in his *Confessions* foreshadowed his later thoughts on the Roman Empire, in the *City of God* (after B.C. 410).[8]

Previously, the pre-Christian Romans used the word *saeculum* to connote a period of time and often linked it with the Roman arcadian games (*ludi lupercales saeculares*). The plural, *saecula*, could designate a span of many or indefinite centuries. In the singular, it was a synonym for *mundus,* "the world." Early Christian authors such as Tertullian (A.D. 160–230) and St.

Cyprian (died A.D. 258), bishop of Carthage, anticipated Augustine's later usage. For the latter, *saeculum* signified frequently a world that is inimical to Christian values, whereas he used *mundus* neutrally to mean the "theatre of human life." Tertullian, an immoderate Christian, eventually left the church to embrace the exclusive, puristic Montanist heresy. He employed the adjective *saecularis* to describe pagan learning (*saecularis litteratura*), which he deprecated. He summed up the dilemma with the famous quip, "What has Athens to do with Jerusalem?" Happily, Augustine and Jerome in the fourth century A.D. endorsed the preservation and interpretation of Roman and Greek thought, for which posterity is immensely in their debt.

Augustine used the word *saeculum* frequently in both the *Confessions* and *The City of God* to designate this age or generation—the age of the human race—but also the future life, the last judgment, or the future end of time or of all centuries past. As a rule, however, he used it in a moral sense: *saeculum* was this world as opposed to the next, or what connects the two— "this pilgrim world." The adjective "secular" described public life—the forum, the stage, sex, gladiatorial games, emperor worship, pagan literature. Even the ancient philosophers embodied the *saeculum* because they dealt "with words rather than realities and are more eager for controversy than for truth." They are fallible in their own way. The "history" he wrote in *The City of God* did not record the pointless politics and wars of the ancient city (the main focus of classical historians), but rather the history of the human race, which for him was the story of salvation.[9]

When Alaric sacked Rome in A.D. 410, public monuments were vandalized and burned, homes were pillaged, and many citizens were killed and their corpses strewn throughout the streets. "Massacres, fires, looting, men murdered and tortured," in Augustine's words. Only the Christian church of SS. Peter and Paul, where refugees fled, was spared by the invaders. Pagans blamed the events on the Christians who had abandoned pagan shrines and condemned pagan worship. Christian mobs had begun to pillage these idolatrous shrines. Christians were blamed not because of the pusillanimity Edward Gibbon attributed to them, but because of their hostility to Rome.[10]

Augustine's reaction differed from the general lamentation. *The City of God* laid the blame for the sack on pagan idolatry, which displeased the Christian God, rather than on disloyal Christians, who had allegedly withdrawn from civic life and refused to engage in pagan worship.

The terrestrial city is mortal; the city of God is eternal. The *saeculum*, although not completely identifiable with Rome, or the terrestrial city, has neither the positive value given it by many today nor the wholly negative value it had for so many contemporary Christian writers. It was simply the unbaptized, unredeemed world—neither an end in itself nor, like the Aristotelian polis, the ultimate site of self-fulfillment. Rome could represent the *saeculum,* from which Augustine was detached. As a Christian, the Rome of rhetoric, of the forum, of the amphitheater, of the Coliseum, of the army, and of the emperors, with their extravagant lives and violent deaths, was not of ultimate value to him.

Nor did Augustine wish to withdraw from that world like a desert monk. Rather, he envisioned Christianity as not only surviving Rome but stepping into its shoes, and converting its citizens as well as its assailants, the Goths. Indeed, modern scholarship has found the church assuming responsibility throughout the empire of those segments of the poorer population who were not citizens.[11]

Thus mankind, as Augustine saw it, was split by two loyalties or two loves: one of the self (*caro*), or the flesh (Augustine's metonym for pride), and the other of the spirit, directed to God. The first, self-love, leads to the "contempt of God"; the second, love of God, to the "contempt of self." Torn between these loyalties, the Christian passes through the world and, with few exceptions, fills the same roles and occupations as did his pagan counterparts. But he does not find complete fulfillment in that world because he is "a citizen of heaven." This dual citizenship, developed in Books XIV and XIX of *The City of God,* was attacked not only by contemporary Romans but centuries later by secularists such as Rousseau, Gibbon, Hegel, and Marx, who found it subversive or alienating.[12]

Although Augustine does not believe in the inherent evil of the *saeculum* (the Manichaean tenet), he does see man's terrestrial existence as transitory (*in hoc peregrinante saeculo*). Not only are individuals afflicted with sin (*anima peccatrice*), sickness, and insecurity, but the city itself is wracked with injustice. The expansion of the Roman Empire gave rise to lamentable internal and external wars. But Augustine found even its peace to be extremely fragile. Only the eternal peace of the city of God was perfect. Until he attained it, a Christian was an exile, subject to secular as well as ecclesiastical authority.[13]

An important qualification of Augustine's bipolar juxtaposition of the two cities was his notion of coexistence (*permixtio*)—the mixing of those in the earthly with those in the heavenly city. Citizens of the earthly city can belong to the city of God, and, conversely, members of the visible church may belong to the earthly city. All depends on which of the two loves they serve the most. The citizen of the earthly city who does not subordinate his body or his pride to God becomes a slave to self-love. But the citizen of the heavenly city, while living like his mundane counterpart, seeks peace within himself by subordinating his body to the will of God. His *dominium* becomes *servitium* even when he is a master.[14]

Vigorously rebutting accusations of incompatibility of the two cities, Augustine wrote to an imperial commissioner, Flavius Marcellinus:

> Let those who say that the doctrine of Christ is incompatible with the State's well-being, give us an army composed of soldiers such as the doctrine of Christ requires them to be; let them give us such subjects, such husbands and wives, such parents and children, such masters and servants, such kings, such judges—in fine, even such tax-payers and tax gatherers—as the Christian religion has taught that men should be, and then let them dare to say that it is adverse to the State's well-being.[15]

Since A.D. 313, Christianity had become an officially tolerated religion (which did not prevent later recriminations, as we have seen), and Augustine made it clear that he did not excuse Christians from most of the duties Romans demanded. His argument was close to St. Paul's and St. Peter's injunction that Christians should obey the emperor and other civil authorities.[16]

Rome could not guarantee man against the vicissitudes of nature, as it had previously been expected to do. Rather, the city of God would take the place of a presumably all-powerful human ruler. Unlike Aristotle, Augustine did not consider the polis the unique matrix of citizenship or virtue. Neither the *saeculum* nor the earthly city was eternal; and neither could guarantee man against chance. Other polities would take Rome's place; it did not really matter which. "As for this mortal life, which ends after a few days' course, what does it matter under whose rule a man lives, being so soon to die, provided that the rulers do not force him to impious and wicked acts?"[17] Such philosophical detachment has been called "subversive" by one twentieth-century scholar, and "indifferent" by another. Are these legitimate characterizations? True, Augustine did equate rulers with "gangs of criminals," but he did so in reference to emperors who "required to play the role of gods, descended to that of beasts," as historian C. N. Cochrane put it. The most Augustine hoped for was that the state would serve as an instrument for the great peregrination, not a final stop. The religions and philosophies of the ancient city had little to offer to this end.[18]

## ii

Augustine took Marcus Varro's *Concerning Philosophy* and counted ninety-six pagan sects somewhat mischievously. The high number reflected pagan dissension, he remarked, as contrasted to the single truth of Christ. What he extracted from this list was that pagans believed that neither *pleasure* (Epicureans) nor *apathia* (Stoic freedom from passion) was the panacea for life. All ancient philosophies, he claimed, concurred that this was "a wretched life," from which man sought to escape. What an astounding confession! The best that ancient philosophers could do was to admit that life was so dreadful that one must flee it (a resort usually associated with Christians)! Neither physical beauty nor friendship was worth living for because both were transient: beauty faded and friendship could sour. Peace was a universal good, but lasted only a lifetime. Honor and position were held in high regard, but these could easily lead to pride.[19]

Augustine's *On Christian Doctrine* was quite critical of classic culture, especially its rhetoric. The fine flow of words left the listener just as uninstructed about their meaning as he was at the outset. "Speaking wisely but not eloquently" would not do either. "He will be of much more use if he can do both." Turning to the substance of pagan learning in a treatise, *Against the Academicians,* Augustine found that pagans could affirm "nothing with certainty, that nothing can be known." Indeed, their skeptical teaching was not only useless but harmful. Augustine harped on one theme throughout: clarity was better than eloquence, since it was useless to speak and not be understood. The critique of rhetoric was perhaps self-criticism, or at least criticism of the profession to which he once belonged. But he did not want Christians to "disregard those human institutions which are of value for their intercourse with Pagans." The Christian should speak wisely *and* well. Interpreters of this fundamental work of Augustine (*On Christian Doctrine*) range from calling it a Christian *paideia* (education, or *bildung*) built on classical learning, to viewing it as a narrow-minded piece of Christian pedagogy. Any interpretation, of course, depends on one's view of Augustine's approach to the two cities. In truth the work was very critical of pagan culture, but not so damning as to forswear its adaptation. Pagan culture was not a value in itself, but could be embraced insofar as it promoted the love of God rather than the love of self. [20]

Augustine never rejected Rome per se because the notion of *permixtio* entailed a certain fusion between pagans and Christians. In truth the latter could be no more certain of salvation than the former. No one, moreover, could expect safe harbor save by the grace of God.

Augustine, in the fifth century A.D., set the limits of the future European secular debate. His clear choice in favor of the city of God would inspire Europe's course for a millennium. During this time, the state was gradually giving way to the church. Already, emperors were abandoning considerable judicial competence in secular cases to bishops. By the time of Gregory the Great (A.D. 540–604), the secular sphere had diminished even further. Gregory no longer felt the need to relate Christianity to secular issues, to pagan learning and institutions, as had Augustine. The age of historians, such as Gregory of Tours (A.D. 538–594) and St. Bede (A.D. 673–735), was the age of God's immanence in nature and time, through miracles and Providence, rather than classical culture.[21]

Much of the learning of the ancient world was lost or known only indirectly through such works as Augustine's *City of God,* Orosius's *Seven Books against the Pagans,* the Epistles of St. Paul, and other books of the Bible and the church fathers. Authentic classical culture succumbed to temporary oblivion. The renaissance of learning did not begin until the Carolingian epoch and did not flourish until the twelfth century A.D. Only when the state began to eclipse the church in the High Middle Ages could one discern the beginnings of a more modern secularism—a Christian secularism.[22]

But back to the issue at hand, Augustine was eschatological only about the Second Coming. He refused to invest too much in the *saeculum,* the empire, or even the church. But he also firmly resisted the theocratic notion of a Christian empire à la Constantine or Theodosius. Insofar as he did not try to baptize the *saeculum,* he was secular. The *saeculum* was to be accepted for what it was—as long as its rulers did not try to "force us into evil." Unlike gnosticism, which considered the world bad, Augustine viewed it as good, because God had created it. Augustine rejected the idea that the *saeculum* could be made into a sacred theocracy in which civil authorities would enforce the law of God, for he was pessimistic about a Christian politics, perhaps the main characteristic of all subsequent Augustinianism. While this was a far cry from the classical "politics of perfection," it had the advantage of not expecting too much from politics—a lesson most modern revolutionaries have ignored.[23]

Certainly many of Augustine's contemporaries—pagans and Christians alike—found in the *City of God* an absolute theologico-political worldview. But Christianity, in Augustine's view, did not deprive man of the city-state; it simply relativized it and transferred many of its functions to the church. The ancient priestly sacrifices were replaced by the sacrifice of the Mass. Care for the poor, the affairs of marriage and education, and so on sooner or later became the church's responsibility. The distinction between the new church and the aged empire was as important as the fourth-century division of the empire between Constantinople and Rome. Capitalizing on dual government, the secular and the sacred mutually limited one another in a cautious embrace that persisted into modern times. As R. A. Markus, an authority on the early church, has shown, Augustine was not enthusiastic about having it otherwise. He saw too many flaws in the state to be triumphalist about an established Christianity. At best he hoped the state would become theologically neutral and leave the church alone.[24]

All this notwithstanding, the temptation to baptize the state and enlist it in the war against the pagan world was strong. The Donatist heretics wanted an untainted clergy and a Christian empire. Augustine fought the Donatists, as he did any inflexible line separating the Christian and the pagan worlds. For Augustine the Christian city consisted of both saints and believing sinners. It would, in any case, be a mistake to preempt heaven by trying to establish a paradise on earth, a sacred city here and now. The state would doubtless still be run by knaves. Far better that it function with relative claims than absolute, Aristotelian self-sufficiency. The church would tend to man's "better part."[25]

The quarrel over the relationship of Christianity to the Roman Empire—begun with Celsus in the second century A.D., continued by Gibbon in the eighteenth century, and protracted by Renan and Nietzsche in the nineteenth century—was that Christian otherworldliness (and by implication its critique of the *saeculum*) sapped Rome of its virility. Yet Clovis, Charlemagne, Joan of Arc, William the Conqueror, Columbus, Catherine of Siena, and Luther all defy such caricature. If anything, most of these Christians were too earthy, too aggressive, too uninhibited. The eighteenth century offered a more valid criticism, namely a Christian intolerance that arose in the fourth and fifth centuries.

Augustine *did* want the church to prosecute heretics, for he believed that only those ultimately belonging to the city of God (as opposed to the *saeculum*) could be saved. His view of persecution was a direct outgrowth of his belief in the absolute truth of Christ that cannot brook contradiction. The uncompromising character of the city of God (wherever it might be) suffused the religious world of the medieval millennium. Augustine advocated prosecution of heresy by the church rather than by the state; he was personally involved in exhorting heretics, such as the Manichaeans, Pelagians, and Donatists, with his sermons and writings rather than by the use of force. *Saeculum,* the key word, marked a cleavage between a perceived world of immoral paganism and a moral otherworldly Christianity. Christian antiquity bequeathed to the medieval world a certain deprecation of the secular world, despite Augustine's concept of *permixtio.*

The coexistence of complete tolerance of all systems and a firm conviction in a body of truth is rare, if it has ever existed. The great difference between the early Christian world and the modern world is that the first accentuated belief in an all-encompassing truth, without an all-encompassing tolerance, whereas the second—at least with its lips—honored an all-encompassing tolerance, because it lacked conviction in an all-encompassing truth.

# Chapter Three
# Thomas Aquinas's Christian Secularism

Thomas Aquinas is essential to a history of secularism, primarily because he drew a famous distinction between what is known by reason and what is known by revelation, concerning the natural and the supernatural worlds. The boundaries of this distinction shifted in favor of reason in subsequent centuries, but the distinction itself dates from the High Middle Ages. The creation of a Christian philosophy that was distinct from theology could arguably be called the origin of modern philosophy. Such a distinction would be matched by secular spheres in politics vis-à-vis the church and in society vis-à-vis the clergy.

Most contemporary philosophy departments do not acknowledge medieval thought as philosophy. It was, they say, largely theological—the work of "schoolmen" or "scholastics"—who were clerics under the tutelage of the church. Not until Descartes, a layman, threw off scholasticism and embraced mathematics, they add, was a truly secular scientific philosophy born.[1]

Many twentieth-century scholars, however, have emphatically rejected this depiction and now believe that the European twelfth century experienced a renaissance that anticipated, if not rivaled, that of Italy after 1300. Cathedral schools began teaching classical texts and the *ars dictamini* (the art of letter writing), which was useful for secular professionals. The clergy was by then mostly literate, the nobility and middle classes becoming so in ever-greater numbers.

France and Italy were taking pride in their newly founded universities. At Bologna, for example, avid law students kept their professors strictly accountable for the curriculum. And one of the greatest minds of the Middle Ages, St. Anselm, taught the ontological proof of the existence of God that Descartes later used in his *Meditations*. In the next generation, Peter Abelard

(1079–1142) stirred up the student world in Paris with his dialectical logic of *sic et non* (yes and no). He argued the heterodoxy that Christian mysteries, like the Trinity, could be found in pre-Christian philosophy. The famous realist-nominalist controversy raged over whether abstract concepts such as beauty and truth have a reality of themselves (like Platonic forms), have only a *basis* in reality, or are simply names (hence nominalism). The debate was Europe's first round of linguistic philosophy. Students flocked to the new centers of learning, not just to enter the clergy but to seek social promotion in the lay world. Social pariahs, such as Moors and Jews, were sought for their superior philosophical and biblical texts.

Illumination arrived in the "flood of translations from Spain, Sicily, and Constantinople" that penetrated northern Europe, posing a challenge of assimilation to the next century. The introduction of Aristotle into Christian Europe was an event of greater intellectual importance than of any other non-biblical author in the history of Europe. It arose due to the Christian reconquest of Spain in the eleventh century. Averroes (1126–1198), Aristotle's principal Spanish–Arabic commentator, stirred up as much controversy in the thirteenth century as Abelard had in the twelfth.[2]

The thought of Aquinas, moreover, insofar as he extricated reason from faith, represents a break with the medieval Augustinian tradition. But to what extent can this philosophy be called secular?

Assuredly, Aristotle was to the High Middle Ages what Newton or Kant was to modern philosophy. It was a token of Aquinas's open-mindedness and of the thirteenth-century intellectuals as a whole that allowed them to accept this enormous corpus so integrally, translate it so completely (through the efforts of Robert Grosseteste and William of Moerbeke [1215–1286]), and assimilate it so eagerly.

It was also a token of Europe's secularity that it could recognize the greatness of a pre-Christian philosophy and find great light in one of its giants. It showed, once and for all, that Christianity had nothing to fear from paganism, which could not be said so confidently of the age of Augustine and Orosius. Thomas never doubted the greatness of Aristotle, whose light, he believed, Christendom was duty-bound to receive.

Yet Aquinas was no slave to Aristotle, revision of whose works was more challenging than simple acceptance. Introduction of the master necessitated careful pruning and adaptation—the work of Aquinas's relatively short life of forty-nine years. Nor was it without intellectual and moral risks, such as the condemnations by the rector of the University of Paris in 1270 and 1277 of scores of Averroistic-Aristotelian propositions, some of which Aquinas had himself maintained. Did he risk the fate of Abelard, who a century earlier made a syncretic use of paganism? Many of the condemned propositions of 1277 dealt precisely with the conflict of reason and faith. Thomas's orthodoxy was, however, vindicated, as was his foray into Aristotelianism, in less than two generations by his canonization as a doctor of the church—indeed as "the angelic doctor."[3]

i

Thomas Aquinas was born in 1225 in his family's castle of Roccasecca between Rome and Naples, where his father, Landolph, was the Count d'Aquino. Thomas was not of Italian stock, but rather German and Norman on his mother's side and Lombard on that of his father, who served the most notorious medieval emperor, Frederick II (1194–1250)—the indefatigable adversary of the Pope.

Thomas was sent to the famous Benedictine Abbey of Monte Cassino when he was five, in the expectation that he would become its abbot. Vocations to the religious life were often viewed as a means to realize worldly ambitions—in Thomas's case, the ambition was to bring the strategic Monte Cassino under the control of the Aquinas family. But Thomas disappointed them. "I will be a preacher," he announced when he was fourteen, hoping to join the Dominicans, one of the recently founded mendicant orders. But then Frederick II attacked Cassino, the monks were dispersed, and Thomas returned home. He left again to study at the University of Naples, where he stayed between 1239 and 1244, finally joining the Dominicans against his family's wishes, and departing on foot for the north via Bologna with the general of his order. Thomas, however, was seized by his brothers and locked up in Roccasecca for a year, during which time a seductress tempted him. But he resisted—his chastity enhancing his intellectual purity—and again set out north. This time he reached Paris successfully, professed his vows at the convent of St. Jacques, and began studies that lasted three years under Albert the Great (died 1280), whom he accompanied him to Cologne. He returned to St. Jacques, where he taught from 1254 to 1256, and wrote the beginning of his *Summa contra Gentiles* (Summa against the Pagans). In the summer of 1259 he returned to Italy, where his genius was recognized by Pope Alexander IV, who kept him in Rome, where he taught at the Curia under three pontiffs and later returned to Paris and Naples.[4]

Thomas was somewhat rotund and exuded placidity. The style of this "dumb ox," as he was called with ironic affection, was calm and confident. The British scholar R. H. Gilbert describes his serenity perceptively:

> Iconography shows St. Thomas calm and sedate, a book on his lap, his fingers expository; he is not proclaiming, denouncing, or wringing his hands.

He was singularly free of the homilists' complaint of living in bad times. Perhaps he lacked the tragic sense. Both the glory that is to come and the present mystery of grace were grounded on physical things; the lowlier they were the better they shadowed divine light.[5]

His enormous powers of concentration enabled him to dictate several of his works to three or four secretaries simultaneously. But he often seemed distracted, as when he was a guest at a banquet hosted by his king—Saint Louis IX. Thinking through the answer to Manichaeanism intensely, and oblivious to his surroundings and the proper etiquette in the presence of royalty, he shouted, "There is the conclusive argument," thumping his fist on the table. In this instance, a writing tablet was quickly provided, for Louis appreciated its renowned metaphysician.[6]

Aquinas challenged much previous thinking in the West. Before the introduction of Aristotle, the prevailing intellectual devise had been St. Anselm's *credo ut intelligam* (I believe that I may understand). Faith could make understanding possible. Aquinas's attitude, however, was that reason leads to faith, which takes over when reason has reached its limit. Faith would then have a rational foundation, and the two would coexist harmoniously in different territories.[7]

Thomas's thought was secular, insofar as it relied on reason; and Christian, insofar as reason agreed with faith; and finally, theological, insofar as his premises were derived from Scripture, revelation, or dogma. The distinction between them and the greatly expanded role given to reason, however, considerably widened the secular sphere noted by Augustine. In fact, Albertus Magnus, Thomas's teacher, had already given the cue: "In science we do not have to investigate how God . . . [works] a miracle . . . but rather what may happen in natural things on the ground of causes inherent in nature." It is a foretaste of the role that "secondary" or natural causes would play in the Galilean revolution three-and-a-half centuries later.[8]

Thomas, for whatever reason, was more deeply optimistic than Augustine about the possibility of understanding the secular world and representing it systematically. He was firmer, more confident, and quieter than the restless, self-reproaching bishop who witnessed the sack of Rome.

One misconception of the history of scholasticism (the philosophy of the so-called schoolmen of the universities studying Aristotle) is that they built a watertight dogmatic system, leaving no room for questions or controversy. The irony is that most Renaissance and Enlightenment intellectuals thought that the Achilles' heel of the scholastics was their dialectical method, their often-contentious public debate.[9]

A burning question of that day was whether the world was eternal, as Aristotle maintained, or had a beginning, or even a creation. The resolution of this question provides a good example of how Thomas related reason to revelation. Aquinas and others could have dismissed Aristotle out of hand, since on the world's origins, he contradicted Genesis. Aquinas believed that where reason and revelation conflicted, revelation, or Genesis, as an inspired text, would prevail. But he also optimistically believed that the two could be reconciled when rightly understood. This could be considered Aquinas's declaration of peace with pagan learning. Both Avicenna and Averroes, the main Spanish Arabic Aristotelians, maintained that the world was eternal, for how could God have created the world "in time," when time would have had to exist "before" the world and without the world? Thomas spoke to this question in his *Summa theologica,* which contains six questions of a "Treatise on the Creation." Only God, he argued, could be eternal, and creation meant bringing being out of nothingness. But creation was not a philosophical question. It was an article of faith, Aquinas acknowledged, "which cannot be proved demonstratively," unlike the existence of God. Thus, we have the two kinds of thinking in two summae, one that could be called secular because it relies essentially on reason and the other theological because it relies on revelation.[10]

Aquinas carefully nuances the distinction between faith and reason in his *Exposition on Boethius' Work on the Trinity:* "The light, which is infused in us by grace, does not destroy the natural light of the human mind . . . it is impossible that that which is divinely communicated to us by faith be contrary to that which by nature is given in us."[11]

Thus, Thomas recognizes two spheres or two ranges of the human intellect: one for natural things, for which divine aid was not necessary, and the other for supernatural, divinely revealed questions. He thought it was good for the unassisted intellect to strive to know God, even though by reason he could only know God's existence, not his nature, for the divine essence is known only indirectly (through revelation).[12]

Thomas is insistent on this point, and his emphasis on the autonomy of the natural (active) intellect breaks with the Augustinian tradition of faith before reason. His accent on the priority of the natural intellect in cognition (in time, not dignity) is certainly more secular than Augustine's.

Both Augustine and Thomas founded intellectual traditions named after them that have survived into modern times. Augustine is a theologian, par excellence, of grace, without which, he believed, no man could be saved. Protestants and Jansenists both favor Augustinianism. The Thomistic tenet—that the individual can perform humanly good actions by his own free will—was adopted by Catholic theologians, particularly at the Council of Trent (1545–1563), and by the Jesuits. The overriding issue between the two is the relative weight given to the human and the divine.[13]

Thomas employs the word *saeculum* in a way similar to some of Augustine's uses of it, denoting an extent of time, a century, a millennium, an eternity—*sub specie aeternitatis* ("from the perspective of eternity") or *in saecula saeculorum* (as in the prayer, "Glory be . . . for ever and ever. Amen"), or even the Last Coming and the Apocalypse. He also uses the word "secular" for "temporal" (that which is measured by time in this world as opposed to timeless eternity) and denoting "earthly," "worldly . . . the opposite of regular, religious, and spiritual."[14]

The *Summa contra Gentiles* is an *apologia* or defense of the faith. In it, references to the newly translated Aristotle, imported from Spain, far outnumber those to medieval theologians. In this, his first major work, Aquinas's aim was to convert Spanish Gentiles, or the Arabic followers of Averroes, by showing the compatibility of Aristotle and Christianity. Aquinas strove to base this work on philosophy alone, although he drew on Christian theology in the second half.[15]

The *Summa theologica* (1265–1272), on the other hand, was a summation of Christian theology, but here, too, important sections of it were based almost entirely on Aristotelian metaphysics and ethics. It differed markedly from Augustine's works, which were more imaginative, Platonist, and indeed often emotional. Thomas's summae, by contrast, were above all orderly. Every chapter of the *Summa theologica* consists of a question, followed by several objections to the question, and then Aquinas's responses to the objections and final answer to the question. This orderly division and subdivision makes issues extremely easy to locate, despite the cumbersomeness of the work as a whole—(the First Part, First Part of the Second Part, Second Part of the Second Part, Third Part). This outline form, introduced by the scholastics, strongly reflects the dialectical didactic then practiced at the University of Paris and other newly founded universities of the thirteenth century.

The Renaissance humanists reacted against an ailing version of scholasticism with felicitous and imaginative prose, but not necessarily superior arguments. Moderns, for their part, have often denied the very existence of something called "Christian philosophy." All medieval philosophy, these critics assert, was really theology. But in fact, it was both—more philosophy in the earlier *Summa* and more theology in the later, which was based on revelation. Both use the deductive method, beginning with a general truth and proceeding to a particular one. The difference between medieval and modern philosophy is pinpointed in the question, what happens when reason and faith conflict? For Thomas, theology, the queen of the sciences, must take precedence over philosophy because of its superior source—it is God's word or revelation—whereas philosophy, theology's handmaid, is based on fallible human reason. By the time of Rousseau, Kant, and Hegel at the end of the Enlightenment, this relationship was exactly reversed. Their modern philosophy does not like the status of handmaid. Still, the fact that Aquinas distinguished between the two, even though he subordinated one to the other, gives philosophy a degree of secular independence that it lacked with St. Augustine and St. Anselm.[16]

Thomas explains this relationship in the *Summa contra Gentiles,* by offering proofs of the existence of God that contain nothing specifically "Christian." Most early modern thinkers, such as Descartes and Locke, also believed that the existence of God could be proved by reason. Aquinas, in fact, exhibited a far greater confidence in the range of reason than most twentieth-century thinkers, who consider the existence of God to be, at best, a matter of faith. Aquinas's contrary conclusion (that God is an object of reason rather than of faith) has enormous intellectual consequences, which the fideist position (that God is an object of faith) lacks. Thomism claims universal, rather than sectarian or private, assent. Necessarily, it brings God into the human and public domain, where previously he had been beyond reason's reach. This is the God of the philosopher: of Plato, Aristotle, Cicero, Rousseau, Franklin, Washington, and Jefferson. It is the God of the public rather than the God of the parish—the God of thinkers rather than the God of the people. It is the *foundation* of faith, not faith itself. And Thomas, not the aforementioned sages, was the thinker who clearly laid down proofs for God's existence. Atheists today, like Averroists then, object to these proofs. Augustinians can also object to them on the grounds that they reduce God to reasonability, thus seemingly denying his inscrutability and transcendence. What is evident is that the existence of God cannot be both a matter of belief and reason simultaneously.[17]

Aquinas, like Aristotle, begins with sense perception, whereas the Platonists, St. Anselm (1033?–1109), and later the Cartesians begin a priori with the mind, with consciousness. For Aquinas, man is "endowed with free will and self movement . . . [he] is master of his actions." Besides having liberty, humans have a rational goal toward which they move. Aquinas believed with Aristotle, and in contrast to us moderns, that there is a natural (teleological) fit between a faculty like reason and its end. He believed that man and the world are consonant. Nonrational animals, Aquinas averred, moved to their goal by instinct and appetite. Only humans had free will to stay on or go off-course. Moderns have compensated for the loss of such an end or *telos* (as in teleology) with *techne* (technology), or skill. Lacking ultimate goals, modern man is seen by his philosophers as muddling along in absurdity.[18]

Augustine had seen communities of monks grow in a somewhat haphazard manner while imperial authority disintegrated. Most often he saw the Christian way of life involve the intermingling of citizens of both "cities": God's and man's. By the sixth century A.D., St. Benedict had founded his monastery at Subiaco, in central Italy, which demanded a more organized separation of the "religious" from the worldly. By the thirteenth century, thousands of houses and monasteries followed different rules, including that of St. Dominic.

Fleeing the world, and entering a monastery, was a very common way of embracing the city of God in the High Middle Ages. Indeed the term "religious" designates persons who take vows and remove themselves from the world to a monastery or convent, as opposed to the "secular clergy," the parish priests and bishops, who live in the world and mix with the laity.

Does Aquinas's endorsement of the life of the religious mean that he sees the world as fundamentally sinful, as Augustine often had? This is a delicate question, for not even Augustine considered the world (*mundum, saeculum*) or flesh (*caro*) an evil in itself—a view that he, in fact, combated. It was man tainted by original sin who used the world and the flesh unworthily and thereby corrupted both.[19]

Thomas definitely believed that grace was not at war with nature, but perfected it. Asceticism or self-denial was a pruning of needless growth, allowing the plant to flourish. The visible world was not bad: it was simply surpassed by a better one. Christ miraculously multiplied portions of bread and wine, and he approved of marriage, thus sanctioning bodily activities. The medieval religious do not condemn the visible world, but subordinate it to a better, invisible one. Of course, when the spiritual and the temporal conflict, the spiritual should take precedence. The secular is good, but it is superseded by the supernatural.[20]

## ii

Pope Nicholas II (1058–1066) instituted a peaceful method for choosing a pope: A college of cardinals would elect him. Shortly thereafter, Gregory VII (1073–1085), probably inspired by the Cluniac monastic reform movement of the tenth century, inveighed successfully against clerical simony (purchase of offices), concubinage (which was widespread), and lay investiture (appointment of bishops by the lay ruler). When these reforms were formalized in 1122 at the Concordat of Worms, the church took on a different aspect. Rome exercised a much tighter control over the clergy, whose lives were no longer so intermingled (*permixtae*) with the laity as were those of the early Christian saints. Even though this separation would not be complete until after the Council of Trent in the sixteenth century, the church of St. Thomas had already segregated its lay and clerical spheres. The separation of lay and clerical and the subordination of the first to the second defined both and gave the laity cause for resentment and revolt. Since the spiritual "sword" dealt with eternity, medieval popes, like Gregory VII, proclaimed papal suzerainty over the entire *saeculum,* including the kingdoms of Spain, Russia, and Hungary. The desired result of such suzerainty would be a united Christendom under one pontiff.[21]

Economic affairs also affected lay-secular distinctions. The clergy was banned from commerce or profit, which were usually tagged as usurious. Tradesmen, for their part, would not for a long time enjoy full respectability and legitimacy in the eyes of the church, which reserved a special place for "God's poor" and the warrior noble. Thomas anticipated the later legitimization of the merchant by granting the profit motive a rank above simple cupidity or greed.

But in doing so, he again distinguished between permissible lay activity and illicit clerical commerce. Secular activities were unworthy of the First Estate, which was supposed primarily to pray. But this did not mean that Thomas endorsed a clerical claim to exclusive "holiness." Nor did it imply any clerical *contemptus mundi,* such as held by Spiritual Franciscans, followers of Joachim di Fiore (1132–1202), who condemned all clerical, even collective, ownership of property. All these extreme views were officially anathematized (almost as rapidly as St. Francis was canonized, two years after his death in 1228, for his exemplary, voluntary poverty). Simply put, the religious were supposed to live a life of detachment from material goods, their personal property becoming the possession of the monastery or convent they entered. The monk left his family no legacy.[22]

Thomas wanted the clergy appointed on criteria other than secular ones. "Respect of persons" should play no role. Whether or not an individual was of a princely or knightly family, like his own, should not be taken into account—only his virtues.

Aquinas did not really have a developed theology of the laity. Those who practice the evangelic precepts (poverty, chastity, and obedience) are religious, and consequently, they live a more dedicated way of life than the laity—and even the "secular clergy." When the clergy do occupy themselves with external things of the world, "they are not seeking anything of the world, but merely [act] for the sake of serving God." Secular life, Thomas continues, "does not dispose one to religious perfection, but is more an obstacle thereto." If the religious take up secular occupations, it should be only for purposes of charity, such as helping widows and orphans, rather than gain. Unlike Augustine, he does not think that the religious need do manual labor, because concupiscence or lust can be curbed by other means. If manual labor is undertaken, it is by "supererogation" or beyond the call of duty. As for the laity, Thomas saw them as living according to *their* station in life, like St. Louis, who strove for perfect detachment from material things yet lived amid regalia. Poverty, like celibacy, was an evangelical counsel, not a commandment. Material possessions were not wrong in themselves.[23]

Thomas would never deny the possibility of lay sanctity or clerical sin. But he insisted that the religious vocation as a state of life was more conducive to sanctity than the lay professions, which he saw weighted down by daily cares. Later, from the Renaissance and the Reformation to the French Revolution and beyond, the subordinate status accorded to laypersons by clerics peaked lay resentment, particularly when the deteriorating morals of clerics themselves became apparent. But Thomas was on solid scriptural grounds for his time. Had not Jesus rebuked Martha for being busy about too many things? Mary had chosen the better (contemplative) part, which would not be taken away from her. Where Augustine's city of God was not divided into two juridical states, but rather by two (intertwined) loves, Thomas found secular and clerical vocations distinguished by dress, dwelling, and canonical status.[24]

### iii

When Aquinas turned from the religious life to government and its relationship to natural law and Christianity, in his *On Kingship to the King of Cyprus,* he had an extensive shelf of classical and medieval sources to refer to: Aristotle's *Politics* and *Nichomachean Ethics,* Avicenna's *De Anima,* Sallust's *Bellum Catalinae,* Cicero's *Tusculum Disputationes,* Josephus's *De Bello Iudaei,* Eusebius's *Chronicorum Libri,* Augustine's *Civitas Dei* and *De Libero Arbitrio,* Isidore of Seville's *Etymologiae,* and the twelfth-century John of Salisbury's *Policraticus,* to mention only those he cited.[25]

Like Augustine, he considered the "common good" far more important than "special interests," either of the king or his subjects. This distinction, sometimes erroneously considered Rousseau's discovery, is as old as political philosophy. The primary common good for Thomas was the peace of the body politic, just as health was a good of the individual.

Louis IX of France (1214–1270) was a model king and Thomas's sovereign. He extended royal justice to all his subjects, hearing petitions in person in the woods of Vincennes. Louis made the first peace with England after the English Plantagenets' invasion of France in the twelfth century. Unlike his grandfather, Philip Augustus (1150–1223), he used good will rather than guile in his dealings with the English. With such a king on the throne, church-state relations were promising. Also, at no other time in European history was there such remarkable flourishing of architecture, sculpture, and stained-glass windows. Witness the Sainte-Chapelle and Notre Dame de Paris.[26]

The questions that had beset medieval kings and clerics alike were whether clerics could be tried in secular courts, be taxed like their secular counterparts, or exercise exclusive control over marriages. Did the clerical role in the coronation of kings suggest that kings were beholden to the church and ultimately to the pope? Could popes dissolve the subjects of their obligation of obedience to their rulers? Could popes dissolve kings of their marital bonds? Thomas, with his customary balance, first opted for obedience to rulers for the sake of peace. But he did see a minimum of political independence from the church when he wrote, "A less holy and less learned man may conduce more [than a pious, but inept ruler] to the common good, on account of worldly authority or activity." Indeed, the papacy had condoned the transfer of power from the Merovingians to the Carolingians on the grounds of the latter's competence and expediency rather than its legitimacy.[27]

Thomas wrote at a time when secular law was pushing forward with alarming strides. Roman law in schools like Bologna uncovered adages such as *lex est quod principui placuit* and *rex est legibus solutus*—that is, "law is what pleases the king" and "the king is not bound by laws." Such sayings, accompanied by the prestige of the Roman Empire, inspired kings to take a more aggressive approach to their feudal, contractual subjects when it was to their advantage. Canon law was developing with equal vigor. Decretalists were ecclesiastical lawyers who pushed the church's advantage in everything from benefices and tithes to marriage taxes. These decretalists, following the lead of the twelfth-century Bolognese jurist Gratian, whose *Decretum* advanced the principle of *plenitudo potestatis,* or fullness of papal power, made the bishops of Rome arguably the most modern rulers of the thirteenth century.[28]

To what extent does one owe obedience to the secular law? Thomas's answers are remarkably clear:

> The secular power is subject to the spiritual, even as the body is subject to the soul. Consequently the judgment is not usurped if the spiritual authority interferes in those temporal matters that are subject to the spiritual authority or which have been committed to the spiritual authority by the temporal authority.

The secular authority is limited from above by the spiritual authority, but it can also be opposed by a limited obedience from below:

> Man is bound to obey secular princes insofar as this is required by the order of justice. Wherefore if the prince's authority is not just but usurped, or if he commands what is unjust, his subjects are not bound to obey him, except perhaps accidentally, in order to avoid scandal or danger.

This is a bold invitation to civil disobedience.[29]

While Aquinas stresses the end of politics, he has little to say about its origin. Like Aristotle, he considers man naturally gregarious, even "a social and political animal." Aquinas not only views spiritual welfare or happiness as the goal of politics but, like the eighteenth-century philosophes, takes into account the physical components of human welfare—geography, climate, diet, health, and population. True happiness, however, is not just Aristotle's virtuous living, but "the possession of God." [30]

When politics fails, Aquinas endorses regicide, but only as a last resort, observing that it may be in the interests of the body politic to suffer tyranny, by which God can purify it and also avoid bloodshed. In extremis, Aquinas is willing to consider regicide if deliberated upon and carried out by public officials, but not by the populace at large.[31]

Aquinas was not the first to legitimize tyrannicide. John of Salisbury had done so in his *Polycraticus* in the twelfth century. Indeed, one eighteenth-century church historian traces the French revolutionary regicide theory to putative medieval origins. But for Aquinas, the real check on royal excess is

> the chief priest, the successor of St. Peter, the Vicar of Christ, the Roman Pontiff. To him all the kings of the Christian People are to be subject as to our Lord Jesus Christ Himself . . . But in the new law there is a higher priesthood, by which men are guided to heavenly goods. Consequently, in the law of Christ, kings must be subject to priests.

This less bloody limitation of royal power from above is preferred to regicide from below. It is a reversal of the pagan and biblical admonition of priestly submission to kings.[32]

The Christian commonwealth is not an end in itself, a polis, the site of man's complete fulfillment. Therefore Christ, a human person and a personal God, transcends the polis. The king cannot be absolute in medieval political theory.

### iv

Aquinas believes that the purpose of government is, as we have seen, to make subjects happy, which can only happen if they are made virtuous. He specifically rejects the goal of power, prosperity, and enrichment as the constitutive elements of happiness, unlike Locke four centuries later. Naturally, Aquinas's kind of happiness assumes a society sufficiently united, religiously or philosophically, to agree to a common definition of virtue.

But can men agree on what brings happiness? Aquinas believes that men always choose what they *think* will make them happy. When men choose evil, Aquinas argues, they are choosing some good that they discern in it. Thus, the thief commits a crime when he steals, but he commits it for the sake of a good he wishes to obtain (money, precious objects, etc.). Man loves iniquity inasmuch as some good is gained through it. "Evil is never loved except under the aspect of good." But man soon discovers that the imagined good does not give him the happiness he desires. Nor do the passions, even though they are good when controlled by reason. Only when he seeks a good that needs no other "needful good" has he gained happiness. (Thus Augustine says, "Our hearts are restless until they rest in Thee.") Nothing else is an end in itself. Anything that exists for the sake of another or exists outside the other cannot be taken as ultimate. Pleasure itself is not sufficient for happiness, but is a *part* of that good, a part not to be rejected, for "an operation cannot be perfectly good, unless there be also pleasure in good."[33]

The only object that satisfies the almost unlimited appetite of man is God, whose happiness exists as his essence. Aquinas acknowledges that man can be united to God in the *vita activa,* but following Aristotle and Christian theologians, he defines happiness, properly speaking, as the contemplation of the good.

But by accommodating the attraction of the senses (the respectable interests of man in the world), by admitting Aristotelian contemplation, and then by introducing the Christian God as the object of that contemplation and the sole, sufficient cause of man's happiness, Aquinas has built a ladder from vice to virtue, reason to faith, action to contemplation, humanity to divinity. (This was made even more explicit by Thomas's Franciscan contemporary, St. Bonaventure [1221–1274], in his *Itinerarium mentis in Deum* [*The Mind's Road to God*].) In short, Thomas has integrated the secular in the sacred, as he did reason and faith.

What determines the goodness of these various ends is reason. Thus, a man must follow his conscience *(synderesis),* even if it dictates evil. But he is responsible for the previous choices that corrupted his conscience. This is a savvy synthesis of moral objectivism (morality remains the same, place to place, time to time, and person to person) and moral subjectivism (morality varies in integrity or purity of conscience from individual to individual, time to time, and place to place). He consoles the moralist, who need not throw up his hands at deviation, which, from Michel Montaigne to the present, has inspired moral relativism.[34]

Aquinas also saw that one way of assuring the prevalence of the good was to repress evil. While Augustine also advocated repression of heresy, Aquinas's instrument for it was for the Christian state to uphold the Christian order by, say, inflicting capital punishment against recalcitrant heretics—something that Augustine shunned but most Christian leaders endorsed before the eighteenth century.[35]

In the Middle Ages, preventing the spread of heresy to the whole body politic took precedence over individual freedom of conscience and speech. The medieval inquisition institutionalized the former; modern liberalism, the latter. Medieval man was bound to follow his conscience, but if in doing so he contravened the laws of society and church, he was subject to punishment for the sake of the common good. The Middle Ages valued doctrinal purity as highly as moderns do physical security. A heretic was as much a menace to the former as a terrorist is to the latter. The former threatened the salvation of eternal souls; the latter, only physical bodies. Only after the loss, or certainly the diminution, of a sense of orthodoxy, did repression of heresy decline and the assertion of rights of the individual become paramount.

Aquinas searched for a harmony among laws: eternal law—the law of God in the universe in its most general form; "natural law"—the moral law known to man through his conscience; divine law—the Ten Commandments; and human, positive, or municipal law—the laws of kingdoms and cities that must not conflict with the higher laws of God and nature.[36]

Natural law originated in the moral philosophy of Plato, Aristotle, and Cicero. It points to a moral law within the nature of man that is knowable by reason, independently of revelation. To take the most obvious example, we know that murder is wrong. We know it independently of God's law, the Ten Commandments, and written (positive) human law. We know it through our conscience. This is another plank in the rational supports of Aquinas's philosophical edifice. It can be called *secular,* as indeed his proofs of the existence of God can be, because they do not rely on faith. Revelation, divine law, *reinforces* natural law (e.g., through the decalogue). In its imperative and majestic fashion, it commands what fallen human nature is loathe to observe. But unlike the Promised Land, the Messiah, or the Incarnation, it did not tell the Jews something they did not already know. Thus, Thomas has raised the secular-sacred distinction, creating another whole domain in Christian philosophy—the domain of everyday morality independent of Scripture and theology. The Protestant reformers, very secular in many respects, upheld the primacy and priority of the Bible to philosophy. Citations of biblical chapters and verse became the guiding light, rather than philosophy.

Did Aquinas therefore believe that man could perform the moral actions necessary to happiness on his own, as *homo sibi relictu?* Augustine, as we know, believed that man could do no good without grace. Aquinas, however, believed that man could, even though man's passions do not obey reason, because his nature is fallen. If Adam's passions had obeyed his reason, there would have been no need for Revelation. But since man does not obey reason, he needs a more powerful motive to obey—namely, God's Revelation in the form of the commandments. This is where the Enlightenment philosophes would part company with Thomistic ethics by arguing that man was either naturally good and rational or that no providence or grace was necessary to help him to do good—only reason.

The record of the twentieth century may have proven that Aquinas and Augustine were closer to the mark than the Enlightenment. Man, "left to himself" has called into question his natural goodness and perfectibility. But does this mean that he cannot do good without grace? Thomas says he can; for instance, he can "build dwellings, plant vineyards and the like." In other words, he can do *natural* good "but cannot produce meritorious works proportionate to everlasting life . . . , for this a higher force is needed, viz; the force of grace." Natural reason, therefore, needs reinforcement by what Thomas calls "the enlightenment of grace."[37]

St. Augustine and St. Thomas identified two approaches to the city of God: Augustine in a not yet completely Christianized environment and Aquinas after a millennium of Christianization. Both described a somewhat bifurcated world—Augustine, the city of God versus the earthly city; and Thomas, a "religious" world of clergy versus a secular laity. Augustine was careful not to identify the visible church with the city of God, the domain of the blessed. Aquinas fully appreciated that the "religious" state was for persons more fully dedicated to God, but justified the lay or secular state and its right to complete happiness. However, justification did not mean equality. Augustine saw mortals living a pilgrim life, scattered amid the two cities. By Thomas's time, the church had become more organized and religious orders were far more numerous. But Thomas had no theology of the sanctification of the laity. In consequence, where Augustine's world was *permixta,* Aquinas's was divided. The two swords within Christendom had become more distinct. Although their sundering lay far in the future, the intellectual seeds of the modern secular world had been broadcast.

# Part II
## Introduction: Renaissance and Reformation

The Early Italian Renaissance continued to express the medieval worldview, but in a different manner and often in a different language. Dante's *Divine Comedy* was written in the vernacular Italian, inspired by medieval troubadour poetry on the one hand and the Latin scholastic theology of Thomas Aquinas on the other. The novelty was the force of his images and the strength of his characters, which transformed dry medieval theology into flesh and blood. Similarly, artists like Giotto and Cimabue painted the same biblical scenes found in stylistic and representational form in medieval cathedral sculpture, but in forceful colors and in perspective. The faith had not changed, but its incarnation had.

Although there is a world of difference between Dante, writing in the first half of the fourteenth century, and Machiavelli, writing two centuries later, both have one thing in common: they reevaluated secular politics at the expense of the church. Dante, in his *De Monarchia*, attempted to make imperial and ecclesiastical authorities separate and equal, subjecting both to himself as poet-judge of the other world in his *Divina Commedia*. Both wrote when the church's *plenitudo potestatis* (fullness of power), both temporal and spiritual, was under attack. Machiavelli, for his part, made a leap over medieval theology, philosophy, and canon law to create a political theory based on personal observation and classical example. Was Machiavelli hostile to the church itself or only to its current corruption? The whole question of Machiavelli's irreligiosity or secularity has recently been reopened. How hostile was he to popes and friars? How completely did his Prince forsake Christian morality and the medieval idea of the ruler in the process? Did he clearly enunciate a modern idea of the state for the first time? And if so, how secular was his conception?

Where the Renaissance was to a great extent a secular movement (at least by the fifteenth century), having religious overtones, the Reformation was a deep and torrential wave of religiosity that swept across northern Europe after 1517—at once a violent protest against Catholic "externalism" and a discovery of the power of the recently translated bibles. The Renaissance affected mostly an elite, whereas the Reformation penetrated far down into lay society and into the vast lake of the European peasantry.

How does a chapter on Luther belong in a book on secularism? Four reasons can be given: (1) he desacralized among his followers the Catholic clergy, beginning with the pope; (2) his doctrine of "the priesthood of all believers" valorized the ordinary Christian; (3) he tipped the balance of church and state in favor of the state; and (4) he fragmented Christendom, inadvertently creating sects that led men to doubt the existence of a true church.

# Chapter Four
# Dante and Lay Secularism of the High Middle Ages

Most people read Dante Alighieri for the same reason they read John Milton: the exhilarating experience of perusing a Christian epic in poetry. *The Divine Comedy* was inspired by the Bible as much as was *Paradise Lost,* but to interpret it Dante drew upon the greatest philosopher and theologian of the Middle Ages, Thomas Aquinas. Where Thomas elucidated human passions in the *Summa theologica* by stating a question, enumerating objections, and giving replies in cerebral syllogisms free of anything sensual or pictorial, Dante used medieval bestiaries to depict human passions. No less a poet than T. S. Eliot has warned against exaggerating Dante's Thomism. Incontinence or lust, for instance, was represented as a wolf "laden with every craving," "a beast without peace," which depicts the insatiable character of that passion more forcefully than any dry proposition. The *Commedia* should be read in Italian, even without understanding the words, if only to experience the beautiful sound of the terza rima.[1]

Dante is considered a secular poet because he was a layman and wrote not in an ecclesiastical language, but in the vernacular Italian he helped create. His sonnets depart from the Gregorian plainchant, by which monks praised God collectively, in that they sang praises to a woman, Beatrice, whose human qualities, he felt, bordered on the divine.

Dante, of course, believed in the Catholic schema of the afterworld—hell, purgatory, and heaven—relegating honored ecclesiastics, even popes, to unexpected "circles." By contrast he praised pagan philosophers and men of letters with awe and was chagrined at not finding them in paradise. Dante's persons are for the most part real, historical figures, rather than allegorical types, such as *Everyman*'s. The poet sets himself apart from medieval theology and philosophy by materializing, individualizing, and historicizing his cast of characters. Ironically, it is a scholastic adage that sums this up best: "Matter is the principle of individuation."

### i

Since Jacob Burckhardt's classic, *The Civilization of the Renaissance in Italy* (1860), the cult of the individual has been seen as the hallmark of the Renaissance. But characterization of the individual was already well developed in the *Commedia*. Indeed, it was the fourteenth-century nominalists (especially William of Ockham) who cast doubt on the reality of universals, such as beauty, truth, and goodness. Dante would not have subscribed to Ockhamism, but his poetry seems to say that general truths are most real when they are expressed individually. He may have drawn from the *fabliaux* (tales) and gargoyle sculpture—those popular, "vulgar" sources of medieval life celebrated by the Russian scholar Mikhail Bakhtin as spontaneous, informal, and physical. Boccaccio and Chaucer would drink from the same well over the next two generations, and Rabelais, two centuries after them.

In his first major work, *Vita Nuova* (1290s), Dante wrote a series of *canzone*, or sonnets written in verses connected by prose narrative. Beatrice had been the object of Dante's devotion since his childhood, even though she may never have existed outside his imagination. "She did not seem to be the daughter of any ordinary man, but rather of a god." He revered her in courtly fashion through most of his life, writing such lines as "Her features were most delicate and perfectly proportioned and, in addition to their beauty, full of such pure loveliness that many thought her an angel." He admitted that he "became a most ardent servitor of love." At such a young age a courtly love poet would say that a lover's "dart" had pierced his heart, causing him intense pain and yearning.[2]

But Beatrice, the *miracolo d'amore*, "died," leaving Dante with an immortal memory. Fascinated with numbers, like all medieval men and women, he wrote of her in Trinitarian terms. The second half of the *Vita Nuova* is a lament over his beloved's untimely death: "Li occhi dolenti per pietà del core / hanno di lagrimar sofferta pena" (The eyes grieving through pity for the heart / while weeping have endured great suffering). Beatrice had ascended to the abode in the outermost circles of the universe as described by Aristotle and Ptolemy. In Dante's words,

> Beatrice has gone home to highest Heaven
> into the peaceful realm where angels live
> For the pure light of her humility
> shone through the heavens with such radiance.[3]

Dante was very much part of the medieval troubadour tradition of love that had originated in southern France 150 years before his time. He mentions fourteen Italian vernaculars, and he drew from many of them to sing praises to Beatrice, to Our Lady, and to God. In this sense the vernaculars, which became Italian, bridged the gap between the profane language of the troubadours' love and the monks' language of divine love. Dante raised to great heights the romance language of *il popolo* to create literary Italian, a language of respectability and extraordinary beauty and greatness.[4]

Was Dante then an Italian troubadour?[5] What sets the *Vita Nuova* apart from the troubadour tradition and that of the twelfth-century cult of the Virgin, which at Chartres and elsewhere exalted womanhood to unprecedented heights, is Dante's intertwining of troubadour and Christian themes. From St. Augustine's *On Christian Doctrine,* St. Aquinas's *Summa theologica,* or St. Bonaventura's *Itinerarium mentis ad Deum,* he could have learned of the natural connection between human and divine love. Bonaventura wrote, "The first source made this visible world so that man, through it as by a mirror and by traces, might be brought to love and praise God the author." The difference between Dante's "new life" and that of the troubadours is that Dante's love does not stop with Beatrice but mounts upward to God. Happiness transcends courtly love, which was "really a love without peace," whereas Dante's attains peace in paradise, where Beatrice is glorified and waits for him.[6]

## ii

The death of Beatrice was not the only misfortune that afflicted Dante in his maturity. The other was exile. When he was thirty-five (1300), after several years of study and several more years of holding political posts in Florence, including that of envoy and rhetorician, he was elected prior of Florence, the equivalent of a city councilor. This was an honor and an opportunity for him to demonstrate his deep civic commitment—a prominent theme in his later writing. He took this commitment as seriously as his religion, and the *Commedia* interweaves the two, such that one can envisage a person being damned or saved on account of his civic record as well as his Christian virtue. The scholar Hans Baron considered this civic accent to be the hallmark of the Florentine Renaissance in the fifteenth century.[7]

Like most of his fellow citizens, Dante belonged to the Guelfs, the party that supported the papacy against the imperial faction; the Ghibellines supported the German emperor, who since the tenth century had claims to Sicily, Naples, and Lombardy. Florentine politics split the Guelf party in two in the 1390s—the Whites (to whom Dante belonged) leaned toward the Ghibellines, and the Blacks leaned more toward the pope. This intertwining of papal and imperial claims led Dante to write *Monarchy,* his one political work, in the second decade of the fourteenth century—the century when the papacy was in Avignon and the imperial restoration in Rome failed.[8]

Florence in the twelfth century had grown wealthy from commerce with the Crusaders in the Levant. The city had developed a powerful textile industry in silks and damask brought from the east, which, after weaving and dyeing, were sold to the German principalities, the Low Countries, and France. Through the straits of Gibraltar, this trade reached England, the Scandinavian countries, and Russia. The population of Florence had quadrupled in the thirteenth century, from some 25,000 to 100,000. Unlike northern Europe with its strong agrarian economies and feudal class, the Florentine nobles frequently lived in the city and built town houses, stimulating Renaissance art and architecture, as well as commerce and industry. The class from which Dante and Beatrice came would develop into that of the Medici, which ruled Florence in the fifteenth century.[9]

Following the very brief pontificate of Celestine V, who resigned the papacy after a few months (Dante condemns him for this "great refusal"), Dante's political career coincided with the pontificate of Boniface VIII (1294–1303). He was an intelligent, witty, and forceful pope, in contrast to Celestine, and contended with an equally forceful king, Philip IV, and with the Colonna family in Rome. Florence, in the middle of these confrontations, was divided between those who supported Boniface and his hope of bringing Florence closer to Rome and away from the German emperor, and those like Dante who opposed him. In 1296 Boniface issued a bull, *Clericos laicos,* which insisted upon the immunity of the church from lay (state) taxation, most notably from that of France. Shortly thereafter, he opposed Philip, who sought to dissolve the Templar Order and appropriate its wealth.[10]

Boniface reasserted eleventh-century claims to papal suzerainty, both temporal and spiritual, in his bull, *Unam Sanctam* (1302). But having grown in power—with increased territory, bureaucracy, armies, and recognition—the northern monarchies put up more resistance to the papacy than had the fragmented German and Italian states. Philip not only rejected Boniface's two bulls but sent an emissary, William of Nogaret, to physically apprehend Boniface, whom he removed from Rome to Anagni. But Boniface was rescued by his supporters and returned to Rome, where the much troubled pope died shortly thereafter on October 11, 1303.[11]

Dante had consistently sided with the White Guelfs of Florence. Marriages in those days were, of course, arranged, and Dante's nuptial with the daughter of a Black Guelf family sometime in the 1290s was meant to patch up political differences, though these persisted. Thus, in spite of his marriage alliance, Dante seems to have participated, while prior, in banishing a number of Black leaders, including his friend, Cavalcanti, who died of malaria in exile. In April 1302 it was the Whites' turn to be banished by the resurgent Blacks, and Dante went into exile, a man without a city, and without much property, which had been confiscated. His four children never knew their father past their childhood. At the time he wrote: "Driven, Dante, from native roof and door / at exile's hands to ply a pilgrim's trade . . . / Wandering the world I've wept two eyes sore / Disdained by death like cripple asking aid."[12]

Dante indicted Florence for corruption and the papacy for simony, greed, and the perversion of the trust of the heavenly keys ("Ah servile Italy, hostel of grief, ship without pilot in great tempest, no mistress of provinces, but brothel!"); his solution was more Ghibelline than Guelf: a classical Rome rejuvenated in the Holy Roman Empire.[13]

The only way to right the wrong was to write about it. Thus, the treatise *Monarchy* addressed a ruling elite that went to the heart of the papal-imperial problem; the *Commedia* (it was not called *Divina Commedia* until the following century) assigned to various "circles" of hell the lustful, the violent, the fraudulent, and the treacherous contemporaries of Florence and of all Europe. Both works depict the existing state of Italy, the *Commedia* more so; both show the alternative radiance of virtue.

His extant letters in Latin all date from this exile and are headed (possibly by an early editor): "Dante Alighieri the Florentine, exiled contrary to his deserts."[14] Dante had his pride. When Florence eventually offered terms for his return—payment of a sum and temporary seclusion—he rejected them out of hand: "Not this the way of return to my country, O my father!" Unless amnesty "hurts not Dante's fair fame and honour," he will not accept it.[15]

### iii

Dante's *Monarchy* was probably written in Latin around 1313, shortly after the coronation of the Emperor Henry VII, who he hoped would restore the imperial authority, eclipsed by the papacy since 1250. It is significant that Dante thinks that a monarchy, and not a republic, would inspire his fifteenth- and sixteenth-century successors. A republic, such as Florence, meant to Dante injustice and ostracism. And a restoration of the empire in Italy raised the age-old questions of precedence regarding pope and emperor.

Scripture supported a distinction between the two powers, such as Christ's "Render unto Caesar what is Caesar's and to God what is God's." Furthermore, when Christ announced that his kingdom was not of this world, he seemed to imply that his church was subject to the emperor in its civil existence. In 325 the Emperor Constantine had convoked the Council of Nicaea to resolve doctrinal (Trinitarian) disputes. With Pope Gelasius (492–496) the distinction between temporal and spiritual powers was clearly drawn, with the spiritual being clearly superior to the temporal, because of the higher realm (eternity) over which it ruled. Constantine bequeathed parts of the empire to the pope in the fourth century, which probably envenomed church-state relations more than anything else in the High Middle Ages. Although the donation was most likely a post-Carolingian forgery, it was used to justify the possession and defense of these lands, and the diplomacy, intrigue, alliances, armies, and wars necessary to retain them.[16]

The church, which had almost successfully emancipated itself from feudal investitures in the eleventh century, became enmeshed in other secular interests in the thirteenth, particularly under Innocent III (1178–1216) and Boniface VIII. But the church not only clashed with the empire over material interests but also laid claims to universal, comprehensive otherworldliness. In his *Unam Sanctam* (1302) Boniface wrote clearly and boldly that there is only one fold and one shepherd and that the two swords, namely, the spiritual and the temporal authority, are in the power of the church. Therefore, "if the temporal power errs, it will be judged by the spiritual power . . . , but if the highest spiritual power errs, it cannot be judged by men, but by God alone." This document, written against the pretensions of Philip IV, could well have been on Dante's mind, as well as the more recent bull, *Pastoralis Cura* (1313) of Clement V, when the poet set out to write *Monarchy*. In the *Commedia* a few years later, he assigned both Boniface and Philip to hell.[17]

Dante thought the election of Henry VII in 1312 as German emperor an opportunity to revive the imperial leadership in Italy, since the pope had been in Avignon since 1309. Later he placed hope in the Franciscans and "Spiritual Franciscans" for a materially pure papacy and a poor church. Thus, the backdrops of *Monarchy* are church-state conflicts, the revival of the empire under Henry VII, and the beginning of the Avignon captivity.

Dante does not dispute Boniface's claim that the church must be "one and holy." Instead he focuses on the temporal monarchy and asks from whom it derives its authority: from God or from his vicar, the pope? His approach to this question is holistic. He resolves the conflict between the spiritual and temporal objectives by taking the word "monarchy" literally, meaning "one rule," a seamless robe. If there is only one rule, there can be no problem of greed, as nothing exists outside the monarch's reach, including the temporal rights of the papacy. A thing, Dante says, "is free when it exists by itself [and it] exists only under a monarch who necessarily unites the good of all," since there is nothing more for him to wish for. Thus, "the well-being of the world" requires a monarchy, Dante concludes, as had Aristotle. The historical, as opposed to the logical proof of this, lies in Augustus's uniting the world at the time of Christ, a far cry from Dante's world.[18]

Book Two of *Monarchy* is dedicated to the proposition that the Romans dominated the ancient world at the time of Christ, not by force but by right. "Right being a good exists first in the mind of God . . . For although the seal [hidden though it is] yields clear knowledge of [the seal] . . . God's will is to be sought through signs." We know God through his effects, as Aquinas stressed. Thus, nothing less than divine will is the source of imperial legitimacy, just as much as of papal legitimacy. Jesus Christ bestowed this legitimacy on the empire by choosing to be born under it and by having his parents, Mary and Joseph, travel from Galilee to Judea to register for a worldwide imperial census ordered by the same Caesar Augustus (actually Tiberius). Furthermore, Christ recognized the legitimacy of Pontius Pilate's governorship by consenting to being condemned under it. Dante's herald for announcing the coming of Christ was the poet Vergil in his fourth eclogue.[19]

Dante's proofs of the supereminence of secular Rome are not, however, purely scriptural. He derives its nobility from Aeneas, survivor of the Trojan War, whose descendants, Dante believed, had founded Rome and whose relatives (Dardanus, Europa, Electra, Africa, Lavinia) spread its rule throughout the world. He alludes to miracles that sanctioned Rome's sway and cites Cicero to the effect that Rome's rule was beneficial to others besides Romans. He also invokes themes that Renaissance and French revolutionary republicans would echo after him: to live for the common good (*res publica*); to live by heroic examples (*exemplum virtutis*); and to die for one's country (*pro patria mori*).[20]

But Dante's solution is not a republican one. He admires Cicero (Tully), Cincinnatus, Brutus, and even Marcus Cato, whose heroic suicide he justifies contra Augustine. That Rome used force to achieve its rule in no way diminishes his approval. In fact, Dante admires imperial Rome the most, because its unity and its sway made the spread of Christianity possible. This was no historical accident, but divine ordination.[21]

After this exposition in Book Two, Dante seems to have arrived at the point where he can only count the emperor's rule as rightful. Yet he surely has no intention of depriving the papacy of legitimacy. Still, having defined monarchy the way he did, it is difficult to see what place the papacy would hold within it. He thoroughly rejects the papal metaphor, which likens the pope to the sun, and the emperor to the moon that reflects only the sun's light. Thus, he dismisses papal arguments for superiority based on the seniority of the papacy to the German Empire, although elsewhere he argues the seniority of the Roman Empire to the papacy. Christ's investing the apostles with the power to bind and to loose individuals does not bestow the papacy with the power to loose and to bind the temporal power. With an Aristotelian rule of distribution, Dante demonstrates that the words "whatsoever thou shalt loose" are not meant to be taken in an absolute sense, that is, applicable to politics, navigation, and so on, but only in a relative sense, that is, applicable to sins. Likewise, the two swords were not meant to apply to church and state, but only to what the apostles would need for their journeys. Finally, regarding the famous Donation of Constantine, Dante does not say it was a forgery, as would Lorenzo Valla (1407–1457), but rather that Constantine could not have given away what was not his to give, since the empire, Dante contends, is inalienable.[22]

Dante is left with two irreducible powers, papacy and empire—in other words, he is exactly where he began. In Aristotelian terms, both differ accidentally or by function: "For the pope's function is one thing and the emperor's another precisely because they are pope and emperor."[23]

Dante's conclusion? There are two powers, and the fountainhead of God flows alike into each. Since temporal authority was not given to the church, the church could not give it (by coronation) to the emperor. To this Dante added a prescription of the Spiritual Franciscans: an extension of the austere rule of St. Francis to the church as a whole, including the papacy. Thus, Dante felt that the church should renounce the temporal kingdom by following Christ, who said that his kingdom was not of this world. If the church were to do this, God's vicar, the pope, would not be able to confer temporal kingdoms, having no title to them. The emperor would receive his mission directly from God.[24]

Dante has used every Ghibeline weapon in his arsenal—history, biblical exegesis, Aristotelian logic and politics, Ciceronian republicanism, and Thomistic theology—to declare the emperor independent, neither subject nor beholden to any earthly power.

Insofar as Dante's vision untied the bonds of the emperor to the visible church authority, it was a secularization of political authority. His reading of Western history to support this relationship could also be considered secular, for he appealed to antiquity, to Vergil's Rome, not to the medieval church councils or decretals. His scorn of decretalists and of canon law, as opposed to Scripture, was characteristic of Renaissance authors, who vaulted the medieval millennium to steep themselves in antiquity. It was arguably the beginning of the fourteenth-century Christian Renaissance. But Dante's classicism, which considered ancient Rome the paragon of virtue, ran side by side with his unbounded admiration for St. Aquinas, as is especially evident in his masterpiece, the *Commedia*.[25]

Thus, pagan antiquity embodied the human virtue of the *homo sibi relictu* (man left to himself, i.e., unbaptized), or the *civitas terrena,* the city of man, which Augustine depicted as living for oneself— for self-satisfaction of all sorts, rather than caritas, or love. For such people a government must exist to punish sin (*poenam peccati*).

Dante pushes this Thomistic rehabilitation of the natural by invidiously comparing the heroes of antiquity to the popes and decretalists of the Middle Ages. This classical bias might have been considered the early modern form of secularism that Machiavelli sharpened. It is shown in Canto XV, where Dante's close friend, Brunetto Latini, is punished with other literati for his excessive concern for *fama,* or reputation. Civic humanism (insofar as it already existed) was clearly a secular value that would get in the way of the glory of God. As Dante scholar Richard Kay writes, "The Inferno is filled with those who sinned to attain fame or fortune that was not their due."[26]

The late medieval nod to secularism was to sacralize the secular power, to make it directly dependent on God (as indeed Luther would make the Christian himself), thus undercutting the authority of the spiritual power on earth, denying it the right to absorb the state, and compromising its universal jurisdiction. This would eventually lead to greater and greater claims of sovereignty on the part of the state. As historian W. H. Reade put it, "A sovereignty of that kind was, in fact, no more than a prelude to the invention of the secular state."[27]

The monarchy's mission, Dante believes, is "to light up the world," without the intermediary spiritual power. This he does in four ways: (l) by glorifying the Roman pedigree of the empire; (2) by tracing its origins immediately to God; (3) by vilifying a millennium of medieval papacy, identifying it with the ignominious decretalist "traditions"; and (4) by negating entirely the Donation of Constantine, thus removing any significance from Pope Leo III's coronation of Charlemagne. Dante thus transfers sacrality from pope to emperor.[28]

### iv

The *Commedia*, which Dante also wrote in the second decade of his exile, most of it probably after the *Monarchy*, was his final judgment of the lay and clerical world. This poem, in which Vergil serves as Dante's first guide, is a Christian pendant to the *Aeneid*. Classical frames of reference are the pagan and rational underpinnings of the Christian supernatural in its *Inferno, Purgatorio,* and *Paradiso*. Dante uses Aristotle almost as much as had Aquinas, and often indirectly, through Aquinas's *summae*. The action of the poem is set fictitiously in 1300, "midway in the journey of our life," when Dante was thirty-five years old and at the crisis of his troubled political career.

Vergil is chosen as the guide through the underworld by Dante's beloved, Beatrice, who in turn awaits Dante at the heights of purgatory to take him to heaven. He is an appropriate guide for Dante, having been the great poet of Rome's foundation by Aeneas, the survivor of Troy. Moreover, it is very important for Dante that Rome be the seat of the Holy Roman emperor as well as of the Roman pontiff. Vergil, the Mantuan poet, represents classical culture; he was a man who went as far as he could by his own powers. He symbolizes human virtue and human reason rather than grace. He reaches the portals of paradise, but no farther.

The journey is didactic: Dante learns in inferno—and we learn with him—how men and women have failed to gain salvation. He is a learner as well as a teacher, a poet as well as an interpreter of God's justice. The journey takes the reader through nine concentric circles of hell to the center of the earth, which is shaped like an inverted cone, and from there up through Mt. Purgatory. It in turn protrudes to the other hemisphere of the earth, to the circles of paradise and the empyrean, the domain of God.

Although rarely if ever enunciated, it is clear that the *Commedia* is Dante's vindication of his unhappy Florentine life and exile. The work of a layman, it is secular in castigating erring clerics as well as laypeople. The first two *cantiche* (as each third of the poem is called) are clearly a righting of wrongs in his personal, civic, religious, and political worlds. At the outset he finds himself "in a dark wood, for the straight way was lost." Dante, like all men, is a sinner, a banished child of Eve. The prayer "Salve Regina" laments man's exile on earth and humbly asks for mercy from a church of sinners. He is not only a political exile from Florence but also a spiritual exile. His path is blocked by three symbolic beasts—the wolf (profligacy), the lion (violence), and the leopard (fraud). Dante may have suffered from the first. Boccaccio, Dante's first biographer, claims that Dante was licentious. Few others mention it.[29]

Dante does not represent sin abstractly, like Aquinas, but humanizes it. He draws on medieval allegories in the *Commedia*, but in an individual and graphic fashion that makes it much more realistic than was common in the Middle Ages (e.g., in the allegories *Roman de la Rose* or *Everyman*).

A key problem of the *Commedia*, as indeed of the whole medieval and Renaissance world, is the assignment of merit or blame to pre-Christian pagans, as opposed to the early Christians. If, according to Christian doctrine, man is born in original sin, which can only be effaced by baptism and which Christ enjoined his followers to administer to "all nations," how can pre-Christians be saved? Yet have not many of them lived meritorious lives? If their merits justify salvation, then are not Christ's life, death, and resurrection unnecessary? Why the redemption? The question would agitate even the Enlightenment (Voltaire and Kant, for example). First, the Christian response was that salvation was a gift, not something one deserved, so no right to salvation could be claimed. Second, Dante introduced virtuous figures of the Old Testament into paradise who had been rescued from hell by Christ's "harrowing" after his resurrection. And, third, he speaks of the antechamber to hell, called "Limbo," where virtuous Roman and Greek pagans, while deprived of the vision of God and true happiness, reside untormented. Nonetheless, it is true that hell is peopled mostly by ancients and paradise mostly by Christians.

How does this fit with Dante's predilection for ancient Romans and his animosity toward so many modern Christians—Florentine Blacks and Whites, and Roman prelates? Does he not seem to prefer the ancients as men *sibi relictu* (left to their own resources): knowledge not faith, valor not hope, eros not caritas? Dante's praise of pagans vis-à-vis his contemporary Christians may, if generalized, reveal a preference for secular Rome over Christian Europe. Let us say he took one of the first steps in the direction that Petrarch continued in the next generation. Historically, and aesthetically, Dante anticipated Petrarch's melancholic wish to have lived in the age of Cicero rather than his own.

Dante's response is one of "great sadness," "for I recognized people of great worth were suspended in that Limbo." But he is a Renaissance prototype who delights in meeting all the pagan and Old Testament heroes, writers, and prophets in this fourth canto. Virgil explains that their fame "wins grace in Heaven which thus advances them." Honor and fame are

Renaissance rather than Christian attributes, as trumpeted in Petrarch's "Letter to Posterity" and Benvenuto Cellini's *Autobiography*.

Dante is stretching Christian grace in the scheme of things to favor the cause of pagans in a heaven, where they can never enter. He sees Aeneas, Caesar, Brutus who expelled the Tarquins, Lucretia, Julia Cornelia, and, "by himself apart, Saladin." Here he greets Homer, "the sovereign poet who like an eagle, soars above the rest," Horace, and Ovid. This is where Dante thinks he belongs from the literary perspective, for he numbers himself "sixth amid so much wisdom," which "diminishes [by] two" when he leaves with Vergil. His self-inclusion is a bit vainglorious from the literary point of view and self-deprecating from the religious one. The philosophers and scientists include Socrates, Democritus, Plato, and Aristotle, "the Master of those who know," so renowned he is not named, and Euclid, Ptolemy, Hippocrates, Galen, Cicero, Seneca, and two medieval Arabs, Avicenna and Averroes. The biblical figures include Adam, Abel, Noah, Moses, Abraham, David, Rachel, and "many others . . . [whom] the mighty one [Christ] made blessed." Dante's placing himself in the first group rather than the second indicates his classical rather than biblical leanings and his civic appreciation of usefulness to humanity rather than the prophetic role of the House of Israel.[30]

In Canto V we encounter the "carnal sinners, who subject reason to desire." Dante's Aristotelianism maintains a hierarchy of intellectual faculties, with the intellect and will at the top and the senses at the bottom. Natural love is never wrong (because it is a blind appetite): it sins by excess or deficiency. Subjecting reason to passion is an irrational act. What is interesting is how those who sin out of passion are punished: they are borne by blasts of wind "hither, thither, downward, and upward." This is the circle of the lovers Paolo and Francesca, who consummated their passion reading the romance of Lancelot. The wind of desire blows them aloft.[31]

This favorite canto spurs Dante's poetic talents. Francesca's acknowledgement of their mutual adultery elicits a sensual and emotional identification with both lovers, as do the best love stories (many of them medieval). Yet Dante simultaneously shows how irrational and therefore evil their infidelity is, not by citing an authority or a commandment, but by painting its futility: the wind of desire bears them aimlessly aloft. They have neither will nor direction.

In the lower circles of hell are those guilty of more serious sins, symbolized by the lion and the leopard—sins of anger and fraud. There too are the Simoniacs, named after Simon Magus, who tried to buy the gift of the Holy Spirit and was rebuked by St. Peter (Acts of the Apostles 8:9–24).

This sin, combatted by the Cluniac and Gregorian reformers of the tenth and eleventh centuries, nevertheless persisted into Dante's time. In the eighth circle a hole is reserved for them to stick their heads in, one Simoniac on top of another; Nicholas III (1277–1280) is there, and also Boniface VIII, whom Vergil addresses:

> Are you already standing there, Bonifazio? . . . Are you so quickly sated with those gains for which you did not fear to take by guile the beautiful Lady [the Church] and then to do her outrage?

After hesitating Boniface answers:

> Know that I was vested with the great mantle; and I was truly a son of the she-bear, so eager to advance the cubs that up there I pursed my gains, and here I purse myself. Beneath my head are the others that preceded me in simony, mashed down and flattened through the fissures of the rock. I shall be thrust down there in my turn when he comes for whom I mistook you when I put my sudden question.[32]

In Purgatorio XVIII, Vergil announces, "As far as reason sees here I can tell you; beyond that wait only for Beatrice, for it is a matter of faith." Dante has switched from descending to ascending—an ascent well expressed by St. Bonaventure's "ladder of love," by which the itinerant soul mounts from reason to faith, from the natural to the supernatural, and from the human to the divine. Awaiting Beatrice after rising from hell to purgatory, we meet the first-century poet Statius, who was being punished for wastefulness, a sin related by inversion to greed. Statius had become a Christian, "but for fear, I was a secret Christian, long making show of paganism, and this lukewarmness made me circle round the fourth circle for more than four centuries." Addressing Vergil he says, "Through you I was a poet, through you a Christian." Statius is a link between Vergil and Beatrice yet expresses Dante's same longing for Vergilian paganism: "To have lived yonder when Vergil lived I would consent to one sun [circle] more than I owe to my coming forth from exile."[33]

Dante's pilgrimage from hell to heaven is illuminated by classical poets. Nothing in his corrupt, contemporary world is equal to the task except Beatrice and St. Francis (and they were both dead). Beatrice and St. Francis can be considered poets—Beatrice by her grace, St. Francis by his universal love.

Vergil's parting words to Dante are: "That sweet fruit which the care of mortals goes seeking on so many branches, this day shall give your hungerings peace." When the glory of Beatrice arrives she says to Dante, "Here . . . you shall be with me forever a citizen of that Rome whereof Christ is Roman." The sentence plays marvelously on Dante's ambiguity between classical and papal, or Christian, Rome. To which Rome does he belong? To the Rome of Vergil, to be sure—classical Rome. But the Rome of Vergil, Dante has been insisting, is the Rome under whose miraculous rule Christ was miraculously born.[34]

We have reached paradise! The radiant St. Francis of Assisi is the perfect pendant to the Simoniac popes. Canonized two years after his death, St. Francis, in the short forty years of his existence, had moved the entire world by love. In his wild youth he had been sensitive to the needs of the poor, and after suffering life-threatening illnesses, he dedicated himself to the life of a mendicant, a wandering preaching friar. He founded an order that practiced poverty literally: the Franciscans wore only a simple brown tunic and a cord. They were given approval by Innocent III in 1214. In 1219 Francis received the stigmata, wounds identical to those of Christ on the cross. Here was something that went beyond Roman valor, *gravitas,* and rationality: an infusion of light and grace that surpassed all human power.

St. Francis, "all seraphic in ardor" and burning with love, holds one of the most honored seats in the empyrean's celestial rose, next to John the Baptist, St. Benedict, and St. Augustine, opposite the Virgin Mary and the harrowed Jewish women of the Old Testament. "Lights" for Dante were divided between the Dominicans and the Franciscans. But he was keenly aware of the pitfalls that religious orders experience. The judgment is put in the mouth of St. Romualdus, a Benedictine reformer:

> My rule remains for waste of paper . . .
> The flesh of mortals is so soft that on earth a good beginning does not last from the springing of the oak to the bearing of the acorn.
> Peter began his fellowship without gold or silver, and I mine with prayer and with fasting, and Francis his with humility; and if you look at the beginning of each, and then look again whither it has strayed, you will see the white changed to dark.[35]

Religious orders run through cycles of greatness, decline, and reform. The thirteenth century was prosperous, and Dante was well aware of the effects. Church bureaucracy and tithe collecting were developing rapidly. Dante delves in a bit of nostalgia as he ruminates that the church of the Renaissance could resemble the church of St. Peter. Grace facilitates virtue, but it cannot prevent sin, or man would not be free.

One reform movement of the thirteenth century, which Dante seems to have admired, was that of the Spiritual Franciscans, who preached an abnormally rigorous observance of St. Francis's rule, especially poverty, as a universal norm for all Christians, including the papacy. Thus, in Canto XII of "Paradise," Dante cites "the Calabrian abbot Joachim [of Fiore], who was endowed with prophetic spirit." The apocalyptic Joachimite renunciation of material wealth became a late medieval heresy that would be used against the Avignon popes in the next two generations, that is, by William of Ockham and Marsilius of Padua. In Dante's paradise, the glowing, seraphic St. Francis carries a heavy sack—the weight of the greedy materialistic Franciscans who gave grist to the spiritualists' heresy.[36]

**v**

Dante's secularism seems to appear at the beginning of a millennial readjustment of the relationship of the spiritual and temporal authority. This cultural upheaval eventually shifted the intellectual and even spiritual leadership of Christendom—from the church and the clergy, where Aquinas thought it belonged, to the laity and the state. Dante's secularism is certainly not modern but, rather, a fourteenth-century poet's Christian celebration of profane love, pagan learning, and imperial authority in the context of Christian redemption. He rescued the secular from Christian oblivion and scorn by showing how much it could be appreciated and be useful to the Christian wayfarer.

The task of reform was continued successfully by a humbler holier soul, Catherine of Siena (1347–1380). She achieved part of what Dante envisioned: the return of the papacy to Rome and its detachment from the politics of France. Unfortunately, the papacy fell back into the trap of Italian city-state politics that would later embroil France and Spain as well. Once again, and to an even greater extent, Rome sank to the level of its contenders. Meanwhile, heresy kept apace, and the attacks against the structure and doctrine of the church reached new levels of audacity. Nothing arrested the momentum of medieval heresy or the Reformation that issued from it. Modern political science was born in the cauldron of papal and Italian city-state politics, which embroiled most of continental Europe by the end of the fifteenth century. Machiavelli, the first modern political theorist, drew his conclusions from this depressing experience with an even greater secular frame of mind.[37]

# Chapter Five
# Machiavelli, Religion, and Politics

Niccolò Machiavelli, it is said, invented secular politics by liberating it from religion and natural law. He abandoned these mainstays of morality, so goes the argument, in order to observe political life as it actually existed, rather than as it should exist. But Machiavelli did have a model—neither Dante's Roman Empire nor the Holy Roman Empire, but the Roman republic of antiquity. His laboratory was political life, but it was learned as much through books (the ten books of Livy) as from the Borgia and the Medici. Did his Renaissance perspective lead him to spurn the whole history of the Catholic Church, or did he hold it in peculiar respect—for its primitive holiness as well as its secular savvy?

i

Between the death of Marsilius of Padua in 1342 and of Machiavelli in 1527, the Italian Renaissance burst forth in pyrotechnic splendor against a gray sky. In its vigorous trecento, quattrocento, and quinquecento periods, it celebrated at once the classic archetype and the modern individual, the intellect and the five senses. The Renaissance "universal man" was interested in everything: grammar, rhetoric, art, history, moral philosophy, theology, mythology, mathematics, magic, and so on. He was much intrigued by the exotic—kabbalah, esotericism, astrology, gnosticism, the Pythagorean and Eleusinian mysteries—rather than by scholastic routine. The humanism of the first half of the fifteenth century asserted civic idealism in opposition to the rising tyrannies and despotisms such as the Sforza in Milan. The Renaissance flourished in Florence just when the republican form of government, strictly speaking, was eclipsed by the arrival of the Medici in 1434.[1]

After the end of the Medici rule in 1492 and Fra Savonarola's apocalyptic republic (1494–1498) came the Florentine republic led by Piero Soderini (1502–1512), in which Machiavelli was a government official.

But the Medicis were restored in 1512, and it was then that Machiavelli wrote his most famous works, *The Prince* and *The Discourses on Livy.* By that time Italy had been invaded three times by the French. Machiavelli was keenly concerned by the liberation of Italy from foreign rule and by the survival of Florence as a free and independent state.

The maxim "The end justifies the means" is believed to be the pith and marrow of Machiavellianism, although this exact phrase cannot be found in his works. This "Machiavellianism" was exported to England, where Henry VIII used it to justify his Act of Supremacy over the English church in 1534 and his subsequent expropriation of monastic lands. A century later it inspired Cardinal Richelieu's raison d'état, which justified Catholic France's intervention on the side of German and Swedish Protestants in the Thirty Years' War in order to weaken the neighboring Catholic Holy Roman Empire. It can be detected again in Louis XIV's expulsion of Protestants (Huguenots) from France in 1685 to complete the religious unification of the French state. A century later, during the Reign of Terror, this logic was seemingly employed to imprison and guillotine thousands of counterrevolutionary "suspects" for the sake of "public safety." Hegel, for his part, believed that the historical dialectic justified the sufferings of the past, and Marx foresaw the triumph of communism as vindicating class conflict and revolution.

ii

The nineteenth century, which had not yet witnessed the worst uses of Machiavellianism, on the whole judged Machiavelli harshly by moral criteria. The twentieth century, which saw the worst atrocities in recorded history, strangely (or maybe not so strangely) came to view Machiavelli in friendlier terms. Perhaps its scholars had become inured to violence by its frequency and magnitude.

Benedetto Croce, the liberal Italian philosopher of history, called for a reevaluation of Machiavelli in 1922 on the basis of his contribution to the construction of the Italian state. Federico Chabod picked up the cue and hailed Croce for having "reversed the age-long denunciation of Machiavelli's 'immorality'." He praised Machiavelli's acute sensibilities and powers of observation that made *The Prince* an integral part of Italian culture and politics. Harvey Mansfield, in the same spirit, tried to place Machiavelli in his sixteenth-century context in order to appreciate his "realism," which was unfettered by Christian moralism. Ancient *virtù,* wrote Mansfield (possibly inspired by Gibbon and Nietzsche more than he realized), had been originally displaced by the weaker Christian virtue after the fall of the Roman Empire. Machiavelli, who was fully aware of this debacle, discovered how that happened—how ancient civilization, specifically Rome with its culture of virility, force, ingenuity, skill, and bravery, was superseded by Christian humility, resignation, altruism, and self-denial. Classical *virtù*, he thought, must be revived from the ruins of Rome and recover its courage and strength for the modern, pagan prince. Mansfield saw classical and modern *virtù* as the antithesis rather than the basis of Christian virtue. The age-old denunciation of Machiavelli's immorality was brushed aside and replaced by acceptance and even praise of that immoralism.[2]

Felix Gilbert, in a 1963 work, *Machiavelli and Guicciardini,* tried to place Machiavelli in the perspective of fifteenth-century humanism, which was definitely classical and Christian, and moralizing as well. Machiavelli was as interested in Christian as in classical virtue, for citizenship consisted in the exercise of both. His political thought, as well as that of his friend Guicciardini, reflected rather than shaped the Renaissance Italy of the quinquecento. If his work is scandalous, it is the fault of the age, which he described rather than invented. His attention to the particular rather than the general can be interpreted as rejection of the opposite Aristotelian-scholastic deductive method.[3]

Quentin Skinner has written the most balanced, researched, and influential history of Renaissance and Reformation political thought in recent years. Unlike Mansfield and Gilbert, who seek to disconnect Machiavelli from Christian Europe, Skinner emphasizes the importance of the twelfth-century liberties of the Lombard communes, which the popes supported against the claims of the Holy Roman emperors. Skinner's view is a much-needed corrective to Hans Baron's almost exclusive focus on the fifteenth century, wherein he found a true civic spirit for the first time in modern history. Still it was not mainly from this Lombardian legacy that Florence and other republics of the Renaissance drew. The Italian scholasticism of writers such as Bartolus of Saxoferrato, Skinner argues, also set standards of the common good that shows considerable continuity with the past.[4]

J. G. A. Pocock stresses Machiavelli's *virtù* as a "drastic experiment in secularization." For Pocock, Machiavelli is the ideal sixteenth-century masterless man "who depends as little as possible on circumstances beyond his control," which meant the mercenaries and their captains, the condottiere. In their place Machiavelli wanted the citizen militia, crucial for both politics and war. Only when a city-state could depend on its own army could it be said to control its own fate. Pocock dwells less on Machiavelli's political unscrupulousness than on his rejection of his contemporary

Savonarola's politics of grace, which was to overcome human weakness and the vagaries of Fortuna. Instead Machiavelli relies wholly on human and secular *virtù* to confront and manipulate the vicissitudes and prevarications of human life. Pocock's Machiavelli acts consistently without any reference to God or the moral law.[5]

Felix Raab, an English historian, finds that the "purely secular rationalization of politics" and its "amazing" conception "as a self-sufficient area of human activity" inspired statesmen like Thomas Cromwell, whom Cardinal Reginald Pole excoriated as the very embodiment of Machiavellianism.[6]

Only one major political philosopher of our time, Leo Strauss, has censured Machiavelli, and that too in the strongest terms, on the basis of a reading of the history of natural rights from the Holocaust back to Plato. Machiavelli, for Strauss, deviated from the Western tradition of natural law—a law of objective morality understandable by reason—that underlay natural rights. Strauss has been treated roughly for his nonconformity.[7]

### iii

Machiavelli, of course, did differ greatly from his medieval and early Renaissance predecessors. Joinville's *Life of St. Louis* (1304–1309) is a good example because it testified to St. Louis's kingship as well as to his saintliness. This patron of monarchs is shown hearing petitions from his subjects, mortifying himself with a hair shirt, crusading for Christ in the Orient, and making peace with England. In *De rege* St. Thomas Aquinas stressed the importance of a king ruling peacefully for the common good, subordinating temporal to spiritual welfare. His disciple, Giles de Rome, toward the end of the thirteenth century, wrote *De regimine principum,* an endorsement of hereditary monarchy (Giles was close to Pope Boniface VIII, and he may have helped author the latter's *Unam Sanctam*). *De regimine* enjoyed wide medieval circulation; a Middle English translation shows his distance from Machiavelli: "Princes scholde knowe moral mater . . . for thereby a prince is taught . . . in what wise he scholde be prince and how he may brynge himself and citeseyns to vertues and good maneres."[8]

Christine de Pizan, daughter of an Avignon courtier, wrote an advice book, in which she, too, bound kings to the moral law. Her *Book of the Body Politik* marshals Aristotle, Cicero, Socrates, and Julius Caesar to underpin her otherworldly conclusion: "So the good prince who desires to reach paradise, as well as glory and praise in the world from all people, will love God and fear God above everything. And he will love the public good of his kingdom or country more than his own as well."[9]

And then there is medieval corporatism. The Pauline image of the mystical body of Christ (1 Cor. 12:12–30) signifies the union and interdependency of all members of the church. The *Policraticus* of John of Salisbury (1115–1180) is the greatest medieval exponent of this organic view translated to the body politic, in which the king is seen as the head, senators as the heart, and peasants as the feet—all equally needed for the body to function properly. The duty of the medieval king is to obey the church, to administer justice, and to keep the peace, war being justified only to bring believers to the faith. Renaissance individualism is largely incompatible with this corporatism, for the Renaissance man abjures subordination.[10]

Two Renaissance humanists of the fifteenth century, Poggio Bracciolini (1380–1459) and Leonardo Bruni (1369–1444), also stressed the moral dimension of politics, but drew more from classical than Christian sources—the humanist practice by definition. Man at the center of the created universe was a favorite Renaissance idea. Humanist concern was with civic duty and liberty; the Ciceronian republic, inspired by idealized antiquity rather than by medieval monarchies; and the Roman republic rather than the Roman Empire, which Dante had admired. Most Renaissance authors did not posit the secular and the sacred as mutually exclusive, or substitute human history for Christian salvation. It was a question of a tilt in direction rather than an about-face.[11]

The transition from the Middle Ages to modernity is also perceptible in the cultivation of an active lifestyle (*vita activa*) by those engaged in commerce, banking (e.g., the Medicis and the Fuggers), entrepreneurial artistry, and patronage (e.g., Benvenuto Cellini and Leonardo da Vinci), as well as in the vicissitudes of politics.

The medieval intellectuals or scholastics had devoted themselves to metaphysics—the study of being or the knowledge of causes, the purposes and ends of cosmic existence; of good and evil; of "universals" and "essences"; and of human wisdom as a reflection of divine wisdom. By contrast the Renaissance humanists substituted the metaphysics of being with knowledge of the self. Some of them unwittingly revived that early Christian heresy, known as Pelagianism, which saw the moral life as largely the result of man's own effort rather than God's grace. The revival of the Sophist—with the tenet "Man is the measure of all things"—spelled a process of secularization, which was a depreciation of monastic submission, found as early as the fourteenth century in the works of Petrarch and Boccaccio. Rather than divine wisdom, someone like Coluccio Salutati

(1331–1406) stressed law, ethics, politics, and economics: wisdom had become prudential. For Bruni, man is a political animal who fulfills himself in the polis rather than in the monastery. While certainly not rejecting the contemplative life completely, the Renaissance humanists treasured the world of the senses and action. It is evident even in artistic renditions of Christ's annunciation and nativity.[12]

### iv

Niccolò Machiavelli was a child of nine when Lorenzo de' Medici, il Magnifico, began his rule (1469–1492), during which Florence was made the cultural, diplomatic, and artistic center of Italy. Two years after Lorenzo's death, in 1492, a series of events began to unfold that would make an indelible mark on Machiavelli's mind, starting with the invasion of Italy by the French. After Charles VIII's invasions of Lombardy and Naples in 1494 and 1495, his successor, Louis XII, conquered Milan in 1499, capturing its governor, Ludovico Sforza. He revisited Italy in 1508, and in 1515/16 Louis's successor, Francis I, in league with Henry VIII of England and with Venice, warred against the pope, the German emperor, Ferdinand of Spain, Milan, Florence, and the Swiss. In the year of Machiavelli's death, 1527, Rome was mercilessly plundered by the army of Charles V—the Hapsburg emperor of Germany, Spain, the Low Countries, Mexico, and Peru. These were tumultuous and unstable years.

Perhaps Machiavelli had been spoiled by a youth spent in the idyllic rule of Lorenzo de' Medici. His description of those years in the *History of Florence* speaks of what he considered superior politics. Lorenzo had quelled the domestic wars of the Italian city-states. His patronage of the letters and public entertainment was famous. Thus Machiavelli tasted the glories of Lorenzo's Florence and saw it vanquished by French armies, and the Medicis exiled. The republic (1494–1512) that followed survived its firebrand preacher-politician Girolamo Savonarola, the 1494 invasion he had predicted, and his castigation of Florence's pleasure-loving ways. In the year of the friar's execution, Machiavelli became defense minister and created thereupon a citizen's militia to replace the mercenaries—a move crucial to his political theory. He went on many diplomatic missions in the 1490s, including one to Cesare Borgia, the pope's son, and another to Louis XII. "He enjoyed much favor from fortune and from the Almighty ... [His] skill, prudence, and fortune" were admired by the world.[13]

The political chaos in which Machiavelli lived barred him from viewing the history of politics as that of Providence, reason, or progress. He could only dream of the expulsion of the "barbarians" (the French and Spanish invaders) and the unification of the peninsula. Otherwise, Machiavelli harkened back to republican Rome, vilifying the Caesar Dante admired and glorifying the republic Caesar destroyed. Only *virtù* and a committed citizenry—a militia instead of mercenaries—taking things into their own hands could ward off similar despotism in the future.[14]

In the first century of the Italian language, Dante and Petrarch spoke of *virtù* in Aristotelian terms, distinguishing moral virtues (prudence, temperance, fortitude, justice) from theological virtues (faith, hope, charity), as, of course, had Aquinas. Virtue was an active love of the good and also a faculty, or power, to do good. Thus Dante exclaimed, "O Splendor of God, give me the power, to say it as I saw it" (the power of a poet). Petrarch went further, recognizing the *vir* of *virtù* to be denoting the virility, courage, force, valor, and the willingness to face death. By contrast *virtù*, for Machiavelli, meant the power, skill, and strength to achieve a political or military objective, not the higher love of man or God.[15]

Machiavelli's prince is a person whose *virtù* was not common. Men are "more prone to evil than good," and the prince must know how to work with, rather than against, human nature. But, above all, he should not feel constricted by religious or moral standards—except to *appear* to observe them. Thus the prince may *seem* virtuous (for this will stand him well in the eyes of his citizens), but he must feel free to do evil if necessary. He must *appear* liberal (generous with money), but he must be careful not to bankrupt himself. "Nor will a wise understanding ever reprove anyone for any extraordinary action that he uses to rule a kingdom or constitute a republic." It is very suitable that "when the deed accuses him, the effect excuses him." This is the closest Machiavelli comes to stating the maxim "The end justifies the means." Does this warrant the portrait of an unscrupulous, de-Christianized secularist? Machiavelli does not say that the action will be "justified," but he implies that public opinion will not blame him for the action if it has a beneficial result. The moral and metaphysical worth of the deed is not broached.[16]

Machiavelli is credited with a modern concept of the state. In the fifteenth century the word *stato* denoted something stable and static, such as a man's station in life, his calling, and his place in a hierarchical order, in French, *état*, as in the three estates of the Estates-General. But then it could also mean a "political command, or *imperio*, over men." This *stato* is something the prince can acquire or lose. It does not refer to an *état* or *stato* within the more medieval, organic concept of the corporate body politic mentioned above. The Renaissance *stato* is the "instrument of exploitation, the mechanism the prince employs to get what he wants." From there it is a short step to the concept of reason of state—*raggione di stato,* or raison d'état. "Reason" gives the state the right to rule over and supersede private interests and conflicts. In the *Discourses* Machiavelli clearly indicates that the prince must act for the common good, even if that entails flouting a prince's private scruples—his reluctance, for example, to shed blood. Romulus founded Rome by "killing his brother [and] deserves pardon; what he did, he did for the common good." Elsewhere Machiavelli states that you cannot judge the results by the means, but the means by the results. Like the difference between traditional and Renaissance notions of *virtù* (i.e., goodness vs. strength), the state is no longer a corporate-cooperative body serving the needs of its members, but an instrument of power and domination, which, Machiavelli will argue, serves better the interests of its citizens.[17]

Compare Machiavelli to his contemporary, the aristocratic Francesco Guicciardini (1483–1540). He too served as a diplomat, mostly for the papacy after 1520. Like Machiavelli he was very anticlerical but, unlike Machiavelli, not very republican. He differed from the Florentine in that he did not think that political experience was as susceptible to rules and maxims. Political reality was too circumstantial, variable, and contingent to lend itself to theory. Moreover, Guicciardini favored the more stable Venetian republic as a model rather than the more volatile Florence. Both statesmen testified to the unscrupulousness of sixteenth-century politics or the existence of a Machiavellianism independent of Machiavelli.[18]

Machiavelli believed that the art of government was the art of war, which, like many other things, was won by *virtù.* Common sense taught that only if a state survived could it claim to have ruled well. Thus defense and conquest (the latter being the best defense) were inseparable from politics. War was as important to Machiavelli as it was to the ancients. It was the calisthenics of a healthy body politic. He would have agreed with Napoleon that peace is only a truce between wars. The scourge of Italian warfare was its use of mercenaries. Machiavelli's advocacy of a citizen militia was probably his most important legacy to early modern political thought. He wanted republicans to fight hand to hand, displaying their *virtù* openly rather than employing artillery or battlements. Romans, like Spartans, were virtuous because they were warriors.[19]

The surprise, however, is that Machiavelli believed that religion could help in waging war. (Many modern generals would agree.) Cicero saw *virtù* in the willingness to face death; in ancient times fear of the gods inspired soldiers to "do their duty," to endure a long siege. "So used well," Machiavelli admits, "religion helped [the Romans] . . . capture [a] city."[20]

In his *History of Florence,* Machiavelli adopts a cyclical view of history, featuring the wheel of fortune. What goes up must come down. "It may be observed that provinces amid the vicissitudes to which they are subject, pass from order into confusion, and afterward recur to a state of order again." They do not proceed in an "even course," he said. The biblical view of linear, progressive fulfillment, ending in the coming of the Messiah or the Second Coming of Christ, is replaced by a classical cycle of human success and failure. Renaissance Fortune is a secularized substitution of divine Providence: history, he believed, is driven more by accident than by reason. Only human *virtù* can wrest something meaningful from Fortune, as long as it pleases her. Unlike Providence, which has an orderly, if inscrutable plan, Fortune is unpredictable and often works at odds with human reason. She steals victory from the jaws of defeat, but also defeat from the jaws of victory. She is, in a word, "fickle."[21]

Oaths sanctioned by religion, Machiavelli believed, are very useful in guaranteeing the loyalty of soldiers to the state. Christianity is superior to pagan superstitions, such as the evil omen of a commander's falling off his horse before a battle begins. Christianity, Machiavelli observes, combats superstition: it "has enlightened their [Roman] minds and dispelled their vain fears." Machiavelli is a long way from the Enlightenment equation of Christianity and superstition. For once, Christianity trumps the Roman religion. But, no matter the religion, it is good as long as it is politically or militarily useful. [22]

Machiavelli's alleged contrast of a meek, pusillanimous Christianity, and of the beatitudes, against Rome's *virtù,* is largely an invention. Christians, after all, were a majority in the army of the late Roman Empire. The ethos of the feudal *Chanson de Roland* is no less military than Christian. Monks have always been a small minority in Christendom and are hardly characteristic of it. Nevertheless they too fought in the feudal centuries.

Machiavelli's contrast between a Christianity, with no place for valor in arms, and Rome and Sparta can only be ironic given his account of Cesare Borgia and Alexander VI.[23] Cesare Borgia in Machiavelli's *The Prince* rose to power by fortune, his father having been elected pope in 1492. But Cesare went beyond papal protection and took matters into his own hands when he assailed the duchy of Urbino and defeated the powerful Orsini and Colonnae families in Rome. He had many other ambitions, such that he "would no longer have depended on the fortune and force of someone else, but on his own power."[24]

Popes could be good warriors, but the *virtù* of the Borgia was used against the other Italian city-states. Far from reverencing the pontificate,

Machiavelli intimates that only by eradicating its overlordship could republicanism establish a strong footing on the local level.[25]

Alexander VI's pontificate began in 1492, the year the first line of the Medici rulers came to an end. Alexander differed from the Renaissance princes and despots only by exceeding their worldliness. Machiavelli deplored his foreign policy, particularly the alliance he forged with Louis XII for the second invasion of Italy. Yet, on the other hand, Alexander excelled in Machiavellian qualities of statecraft, possibly more than any other contemporary save his son. Machiavelli thought that Alexander "showed how far a pope could prevail with money and forces." Between a "bad" Renaissance pope and a fanatical republican friar, there was not a doubt whom Machiavelli preferred. After Alexander, temporal potentates trembled before the Vatican. Even after the fall of the Medicis, the papacy emerged as the real victor of the Italian wars in the person of Julius II (1503–1513), "the warrior Pope," who pursued the aims of Alexander successfully. He left to Leo X, his Medici successor, "this pontificate most powerful."[26]

Machiavelli was perhaps the first political thinker since Plato to stress the distinction between *être* and *paraître,* knowledge and reality, and semblance and public opinion. Plato believed that only the philosophers could see the distinction; the mass of men were taken in by *doxa*—that is, opinion based on appearances—and by the shadows on the wall of the cave. Michel Foucault stressed this early modern *epistemé*, or theory of "representation," as opposed to Aristotle's concern for "being."

v

Chapters 16–19 and 21 in *The Prince* constitute the essence of what most people think of Machiavellianism: immorality, deceitfulness, unscrupulousness. It is frequently excused on several grounds: (1) Machiavelli wrote *The Prince* tongue-in-cheek, and he did not mean it to be taken seriously; (2) he did not invent the behavior recorded in *The Prince* and *Discourses,* but simply observed it in ancient Romans and in his contemporaries, including the popes; (3) his maxims *work* even though they are morally wrong and truth should cede to pragmatism; and (4) the Christian virtues and beatitudes do *not* work in the harsh world of politics.

Machiavelli seemed to set up a double-entry register for his prince. In one column are all the deeds, statements, and Christian virtues that gain him public credit. In the other, one finds the unscrupulous, outrageous machinations that underlie those apparently virtuous acts he "prudently" condones. The prince's greatest breach of faith is with the body politic. He lives by the Averroistic double truth: one for the populace and another for the philosophers.[27]

In *The Prince,* for example, he writes of the great danger of keeping envious men in one's service. Agathocles (B.C. 361?–289), master of Syracuse, Machiavelli recounts, enjoyed a reputation of kindness and *virtù* of mind and body. As praetor,

> he assembled the people and the Senate of Syracuse as if he had to decide things pertinent to the republic. At a signal, he had all the senators and the richest of the people killed by his soldiers. Once they were dead, he seized and held the principate of that city without any civil controversy.

Is this the *virtù* Machiavelli recommends—killing one's fellow citizens, "without faith, without mercy, without religion . . . to acquire empire"?[28]

A "modern" counterpart was the case of Liverotto of Fermo, the first soldier in the army of Paolo Vitelli. With the consent of Fermo's citizens, he rode into Fermo with a hundred horsemen. He threw a banquet, gave a speech lauding Alexander VI and Cesare Borgia, and ushered his uncle and other potential rivals into a room where they were all killed. Liverotto became recognized as prince, or more accurately as despot, because of the fear he inspired. He ruled Fermo safely for a time, proving Machiavelli's principle that it is better to be feared than to be loved, but he too came to an end appropriate to his beginning. He was strangled by Cesare Borgia.[29]

Are Agathocles' and Liverotto's murders examples of *virtù?* Are these tyrants acting according to *raison d'état* and therefore justified? Machiavelli equivocates on this matter. If they are executed "at a stroke" to secure power, and the agony is not prolonged (how considerate!), the harm is limited. "Timidity" and "bad counsel" would have been the downfall of the murderers. In the end Machiavelli's moral sensibilities are overcome by reasons of state. Thus he states that the perpetrators can "have some remedy for their state with God and with men," since their crime is contained. One could consider Machiavelli "a citizen of the two cities!"

These are not isolated examples. Machiavelli epitomizes: if a prince becomes a patron of a city formerly free, and "does not destroy it, he should expect to be destroyed by it." Thus Cesare Borgia did well to "eliminate the Orsini chiefs, since he had dispersed those of the Colonna house." These two Roman families traditionally gave the pope the most trouble.[30]

Chapters 7 and 8 of *The Prince* show Machiavelli at his most sinister. Clearly he intended the lion, representing the prince, to do a great deal more than just roar. He had also to know when to devour. Where in the *Inferno* Dante put such malefactors deep in hell for sins of the will, Machiavelli considers their actions examples of prudence. But even when such actions are covered with the veil of virtue, which Machiavelli recommends, one finds this veil frequently torn off and the perpetrator overthrown.

A last example is apropos—the Pazzi conspiracy against the Medicis in 1478, the year of Lorenzo de' Medici's accession to power that Machiavelli recounts in *The History of Florence*. The conspiracy was sparked by the envy of the Pazzi for the Medici family, aggravated by personal injuries and insults, such as an unfavorable inheritance decision and disappointment over a dowry. At the root of this clash was a struggle for family, honor, and power—three forces that drove the Florentine public life.

The Pazzis enlarged their circle of conspirators to include the head of the papal forces (who eventually refused, because of the sacrilegious setting of the plot), and the archbishop of Pisa, a rival of the Medicis (who apparently had no such scruples). After two failures the conspirators chose a mass at the Cathedral of Santa Reparata as the scene for the crime. Lorenzo managed to defend himself from the stabbing, which started just as the priest was consuming the host, but Giuliano de' Medici was mortally injured, whereupon the crowd rallied behind the Medicis and massacred the Pazzis and their followers, brandishing their body parts in the streets of Florence.[31]

Machiavelli neither condones nor condemns this atrocity. He narrates it. He cannot approve it, because it failed. Moreover, he was an admirer and a protégé of the Medicis at the time he wrote his *History*. Since it was an occasion for Machiavelli to show his humanist sentiments and deplore the Pazzis for this atrocity, he puts a remarkable eulogy of Giuliano in the mouth of a survivor, Lorenzo de' Medici. But had the conspirators been the Medicis, would he have condoned it?[32]

### vi

Machiavelli's *History of Florence,* commissioned by a future Medici pope, Clement VII, was left unfinished when Machiavelli died in 1527. While the book is characteristically frank and critical of the church's temporal power, a passage on the early church strikes a different chord:

> In these times the popes began to acquire greater temporal authority than they had previously possessed; although the immediate successors of St. Peter were more reverenced for the holiness of their lives, and the miracles which they performed; and their example so greatly extended the Christian religion,

that princes of other states embraced it, in order to obviate the confusion which prevailed at that period. The emperor having become a Christian and returned to Constantinople, it followed, as was remarked at the commencement of the book, that the Roman empire was the more easily ruined, and the church more rapidly increased her authority. Nevertheless, the whole of Italy, being subject either to the emperors or the kings till the coming of the Lombards, the popes never acquired any greater authority than what reverence for their habits and doctrine gave them. In other respects they obeyed the emperors or kings; officiated for them in their affairs, as ministers or agents, and were even sometimes put to death by them.[33]

This passage reflects a conventional humanist view of the early Christian church—a church far purer than the Renaissance church. Although this early church acquired temporal possessions, its authority stemmed from its holiness, revered habits, pure doctrine, and brave martyrs—an image also celebrated by Erasmus and the Protestant reformers. Moreover, according to Machiavelli, these popes remained submissive to secular authority—a key reason for his approval. But did Machiavelli really admire this holy church in itself, or did he merely use it as a foil?

Another passage, from *The Discourses,* might shed light on the question. Just after discussing Roman examples of terror and virtue as alternate ways of bringing a republic back to its original virtue, Machiavelli digresses on the medieval mendicant orders and their relation to primitive Christianity:

But as to sects, these renewals are also seen to be necessary by the example of our religion, which would be altogether eliminated if it had not been drawn back towards its beginning by Saint Francis and Saint Dominick. For with poverty and with the example of the life of Christ they brought back into the minds of what men had already eliminated there. Their new orders were so powerful that they are the cause that the dishonesty of the prelates and of the heads of the religion do not ruin it. Living still in poverty and having so much credit with peoples in confessions and sermons, they give them to understand that it is evil to say evil of evil and that it is good to live under obedience to them and, if they make an error, to leave them for God to punish. So [the people] do the worst they can because they do not fear the punishment that they do not see and do not believe. This renewal, therefore, has maintained and maintains this religion.[34]

This passage is less conventional than the preceding one. It was not written for a Medici cardinal. Composed two centuries after Dante's *Inferno,* it bears out one of Machiavelli's main ideas: the importance of bringing institutions or "orders" back to their foundations. Machiavelli here seems to praise orthodoxy, poverty, confessions, and obedience to the church. The penultimate sentence suggests the difficulty of the task. That Machiavelli is willing to criticize the hierarchy in the same paragraph as he praises the mendicants makes his encomium sound sincere. The last sentence, however, seems to indicate that he esteems them for their order and power, which alone keeps the papacy from undoing them. Piety is useful in religion as in battle in keeping the enemy at bay.

These quotations contrast, nonetheless, with the author's advice in *The Prince* to practice evil because most others practice evil. The early church fathers, the Franciscans, and the Dominicans represent a certain oasis of good in an evil world.

### vii

When clerical prestige, added to secular virtue, is damaged, the corruption is perverse and potent. Machiavelli witnessed the secularization of political theory in church and state. His anticlericalism, however, is somewhat reassuring because it suggests he held the church up to higher standards than did some of its leaders (as cynics frequently do). If he was resigned to the secularization of Italian politics, he was not resigned to the corruption of church doctrine. If he tolerated political immorality, he did not tolerate heterodoxy. In fact he was much less "heretical," strictly speaking, than William of Ockham and Marsilius of Padua in the fourteenth century who would have forced "supreme poverty" on a pontiff they called a "heretic," namely, Pope John XXII. Machiavelli did compartmentalize politics and religion, as many Italians still do. He acknowledged a divorce between politics and morality. He respected what *is* rather than what *ought to be*. He never invoked natural law; instead, like a social scientist, he looked upon the world as a living laboratory—something to be observed and reduced to maxims. He and Guicciardini declared the independence of politics and the chasm between morality and the church. As such the church was a political entity in the secular camp.

Machiavelli saw this world of secular politics as fairly irrational—a furious maelstrom sucking in everyone. Thus the heads of city-states, of papal and imperial monarchies, were tragic heroes, who were not entirely responsible for their actions or the fate of their states. The maintenance of the papal states snarled the papacy in a diplomatic and military mesh. Only those who separated themselves from the world—early martyrs and later friars—could escape from its eddy. But Savonarola's death demonstrated that not even monks were immune. Machiavelli had more of an Augustinian view of church and state than a scholastic one. Augustine thought that states were run by criminals; Machiavelli proved it. Both saw some incompatibility between the city of God and the terrestrial city. But each embraced a different one. The seventeenth-century Augustinians—the Jansenists—would pull faith and politics apart. So did Machiavelli, but, unlike them, he embraced politics rather than faith. Machiavelli's war was about power, and the only moral regulating it was victory, no matter the means. Many European statesmen would adopt precisely this approach. Francis I's alliance with the Turkish sultan Suleiman I in 1518 against the Christian Hapsburgs and Henry IV's probably apocryphal 1593 statement explaining his conversion—"Paris is worth a Mass"—could be considered examples of Machiavellianism, or raison d'état, whether or not they originated in a reading of *The Prince,* whose maxims spread further than the book.

Machiavelli's republic was the ultimate meaningful association for a citizen. It allowed him to participate in, fight for, and determine the course of his life, and to control chance, not through contemplation but through ceaseless observing, reasoning, and acting. Machiavelli was the first to identify politics as the pure and simple struggle for power. He believed that if you took his realism seriously, you would not be deceived; you would possess a strategy of survival and a method of serving the public good.

Like Marx and Nietzsche, Machiavelli is one of the West's great demaskers and secularizers. Whereas Marx and Nietzsche sought to annul religion, Machiavelli sought the source of its power, to enlist it for secular purposes. Something need not be good in itself as long as it is good for the state. The whole problem here is, when does this logic become unreasonable, immoral, and despotic?

Who is Machiavelli's prince? He is the Renaissance masterless man, responsible only to himself and obedient only to the laws of his own creation. Benevolent when convenient, cruel when challenged, pious when useful, conspirator, murderer, assassin when necessary, populist, republican or despot as called for, a man fixated on one thing—power and war—employed for the good of the state, which is inconceivable without him, Machiavelli's prince subsumes the state in his person. Cosimo, Lorenzo, Savonarola, Alexander VI, Cesar Borgia, Ludovico Sforza, Louis XII are his names. These individual portraits are realistic and colorful. Their very vices account for their posthumous popularity. They cannot be found in the medieval, stylized mirrors of kings and their generic, colorless precepts.

Yet there were quandaries. Machiavelli's individualistic prince was bound by many maxims. How can the prince be a masterless man when he, Machiavelli, is clearly his master who imposes rules on his creature? Is the prince really the autonomous man usually associated with Rousseau and

Kant? Or a stock creation of the Renaissance, like the courtier of Baldassare Castiglione? The most shocking thing about Machiavelli was his benediction of evil, which Leo Strauss boldly branded "diabolical."[35]

Whatever Machiavelli was, he was not lukewarm, as is evident from his approval of Agathocles' and Liverotto's assassinations. But he did not banish religion from politics. Indeed he created a role for it—to serve the ends of the state. As such he overturned the putative medieval subjection of the realm to the pope and to God and inaugurated the secular political theory of the modern world.

# Chapter Six
# Luther's Centrifugal Reformation

### i

Luther's secular spirit is paradoxical, because it is clothed in religion and visible more in its consequences than in its original thrust. The Reformation was the work of a Saxon monk, who traveled 1,500 miles from Erfurt to Rome and back in 1510 to plead a case for his Augustinian order before the Curia. It was not yet the Rome of St. Peter's basilica; nor did it repulse him. Rather he was troubled inwardly by the great problem of the approaching age: "How can I be saved?" Indubitably, this was a quintessentially religious problem. The solutions Luther reached, however, had prominent secular dimensions when they led him to reject, step by step, Catholic tradition and authority.

Rome did not seem to answer this problem of hope. Nor did his monastic training in Erfurt. He was already preoccupied by the devil, as was so common in this "waning of the middle ages." By the end of the century this obsession beset whole nunneries and villages. Some have interpreted Luther's outburst in the Erfurt choir—where he blurted out "ich bin nicht"—as a renunciation of diabolical temptation; others, as a denial of obedience to his father, who opposed his claustration and ordination. His spiritual malady was scrupulosity and guilt, which his confessions did not assuage. He was a professor of theology—a learned one who, around 1513–1515, while lecturing on the psalms and St. Paul's Epistle to the Romans, hit on a statement—"The just shall live by faith"—that became the centerpiece of his moral theology. Man doesn't justify himself—by his efforts, his "works," his successes. God imputes righteousness to an unrighteous, undeserving sinner. (The whole of Lutheranism, its internalization of guilt, its passive redemption—lies here.)[1] By contrast its antidote—Tridentine Catholicism—stresses free will, "spiritual exercises," active works as well as faith and grace.[2]

The Lutheran Reformation was more of a spiritual than a secular movement, emphasizing inwardness that became the core of subsequent German religiosity. But in revolting against Rome, Luther broke up the unity of Christendom, which led to sectarianism and ultimately to doubt as to what was "the true church." Luther's attack on the Catholic conceptions of priesthood, celibacy, monasticism, marriage, and free will (all categorized as "works" in his tracts) certainly undercut their supernatural value and so secularized them. Above all, Luther changed the relationship of the clergy to the laity by denying the supernatural powers of the priest. Furthermore, on key issues, he was challenged by close and distant followers: images, the Blessed Virgin, the Real Presence, political authority, and violence and war. The result was a splintering and fragmentation of Christianity such that Luther was fighting a ubiquitous war against the enthusiasts of Reformation—Carlstadt, Muenzter, and Zwingli—as well as Catholic counterreform.

The form in which the Reformation reached England was not initially subversive, for the Tudors kept it under control. Under the Stuarts, however, the Calvinist (Puritan) "saint"-revolutionaries believed that the beheading of a king in 1649 (Charles I) was an act of piety and righteousness. The aftermath of the English civil wars (the Stuart Restoration, 1660–1689) produced another solution at the hands of John Locke: religion should be relegated to individual consciences. The Reformation has been called *theonomous* rather than secular—a vast inward search for God. But in attacking the linchpin of medieval Christendom, the papacy, and monastic asceticism, Luther opened the floodgates of private judgment, scriptural strife, the right to resistance, and, paradoxically, claims to untrammeled state sovereignty.

Luther objected less to the crass sale of indulgences by the Dominican friar Tetzel in 1517 than he did to its theological and religious implications: that the pope had the power to remit the penalty of sin, even of souls in purgatory, and that living Christians should be given such an easy opportunity to pay those penalties (not the sin itself) by the simple purchase of a piece of paper, rather than by a deep, personal penance.

At first the conflict drew little attention from Rome, which regarded it as the affair of another crazy monk. There had been many medieval heresies: Spiritual Franciscans, Hussites, Lollards, Cathars, and Waldensians. All of them had been targeted for suppression. Certainly no one had foreseen the magnitude of what Luther would produce. Emissaries came forth to deal with him on the issue of indulgences—Professor John Eck, Sylvester Prierias, and Cardinal Cajetan. These encounters quickly metastasized into Luther's full-blown challenge of papal authority.

The format, or venue, was the debate, the diet, or the colloquy—medieval scholastic institutions for the most part, in which both sides were heard but from which agreement did not emerge. The issue became quickly apparent—not the indulgences, but whether the pope was inerrant. Luther contended that "papal decretals occasionally are erroneous and militate against the Holy Scriptures" and that the Holy See is not apostolically founded or placed above other sees. At the Leipzig debate in 1519 where Luther encountered John Eck, only fourteen months after the posting of the 95 theses, Luther took up again his notions of true penance not to be substituted by indulgences. He defined a Pelagian as anyone "who maintains that a good work and a penance begin with the hatred of sins."[3]

The indulgence issue, as can be seen, permuted into a question of authority that underlay the secular substance of Lutheranism. Luther contended that indulgences were without scriptural foundation. Rome maintained that it had the authority to do what it did without basing its authority on anything other than "Thou art Peter." Everything should be ultimately referred to that scriptural passage, not to every challenger's interpretation of other passages. Luther was excommunicated in January 1521. Three months later, at the Diet of Worms, he was convoked and given safe passage from Saxony to Worms. He refused to repudiate his writings, unless refuted reasonably by Scripture—prompting the famous words, "Here I stand." The cord linking Luther to Rome, which had been unraveling for months, broke. Although he was given safe passage back to Wittenberg, he was not entirely safe there, because he was under an imperial ban at this point. His elector, Frederick, let him wall himself up in a nearby castle at Wartburg.

## ii

Already, in his *Address to the Nobility of the German Nation* (August 1520), Luther had appealed to a nascent nationalism of Germany versus Rome when he laid siege to the three walls that the "Romanists" had raised to maintain their spiritual hegemony (that the temporal power has no jurisdiction over the spiritual, but rather the laity is included in the spiritual estate, thus redefining the church; that no one may interpret the scriptures but the pope; that no one may call a council but the pope). Knocking down those walls successively would secularize church-state relations in Germany. First, papal jurisdiction over Germany would cease; second, the pope's dominion over Christian doctrine (scripture and councils) would be overthrown; third, the pope would be greatly reduced in his authority over the church itself. Its sway no longer over Christendom, the church would no longer be universal. Instead, national or territorial churches would take its place here and there—the ones responsible to local secular authorities rather than to an entity catholic in space and time. The attack on Rome, minimized by ecumenical historians of the Reformation, is bound up with Luther's more positive efforts to reform. These have little to do with simony, benefices, Peter's pence, and clerical concubinage; they relate to the inner spiritual life, the core of Luther's religion. For Luther, man is saved by the grace of God, *sola gratia*. The Roman religion—pilgrimages, rosaries, celibacy, vows—was derided in Erasmus's *Praise of Folly* (1509), and Luther also condemned all its rituals in one fell swoop as "works" by which Christians felt they merited heaven. If works do not save, none of these is worthwhile.[4]

The *Babylonian Captivity of the Church* (October 1520) on the surface alleged notorious papal worldliness to strip it of its legitimacy: "The Papacy is truly the kingdom of Babylon and of the very Anti-Christ." But the essential is the application of *sola scriptura* to the sacraments. Of the seven, only two would remain in Lutheranism ("Only two can be proved from the Scriptures"), communion and baptism (and for the time being penance). But if Luther considered the Eucharist sacred, it was not "too sacred to be delivered to the common people." Another wall—that dividing the two classes of Christians, the clergy vis-à-vis the laity—had been knocked down, clearing the way for secularization.[5]

Luther's *Freedom of a Christian* was published in November 1520, weeks before the first effort for his excommunication by Leo X, which could not have come as a surprise—not only because of Luther's frequent denunciations of the pope as "the Anti-Christ" but also because of Luther's rejection of Catholic doctrines, which he denounced truculently in his three inflammatory tracts of 1520. His view of Christian freedom for instance was quite arresting: "A Christian is a perfectly free lord of all, subject to none . . . A Christian is a perfectly dutiful servant of all, subject to all." The maxim captures Lutheran paradoxality. Man lives under the law but is set free by the spirit. "Love God and do as you will," as St. Augustine had said. Man performs works, says Luther, yet is not justified by them but by faith. His sense of guilt is the basis of his righteousness and justification. He accepts authority, but his spirit is free. He is a very physical, even vulgar, being; this does not hinder his election. "Not only are we the freest of kings, we are also priests forever, which is far more excellent than being kings." Here Luther introduces the concept of "priesthood of all believers," by which he exceeds the attack on clerical superiority, in *The Babylonian Captivity,* and seeks to liberate priests from Rome and the layman from clergy: "We are all consecrated priests through Baptism . . . Christians are truly of the spiritual estate."[6]

Consequently Luther did not recognize the Catholic Sacrament of Holy Orders, which established an indelible mark in the soul of the priest, setting him apart and above the laity. In the two sacraments he did retain, the Holy Eucharist and baptism, he insisted that the priest does not have the power to perform the miracle of transubstantiation, the changing of bread and wine into the body and blood of Christ (while the accidents of bread and wine subsist). Luther believed that Christ was cosubstantial in the host—a supernatural, "real presence"—but was not brought about by the "work" of the priest. Luther's version represents a diminution of God's presence in the world (the sacrament is not preserved in permanence in the tabernacle)—a desacralization that Zwingli's, Calvin's, and the sacramentalists' purely symbolic interpretations of the host augmented. Medieval man found God immanent in nature, in miracles, and in the sacraments; Protestantism tended to desacralize nature and replace the supernatural, transcendental character of the liturgy by the written and oral word.[7]

Throughout traditional Europe, thousands of masses were celebrated daily by ordained priests. These masses were defined as "sacrifices" whereby Christ was sacrificed daily and was daily offered up in a real, although unbloody, manner, but as effectively as he had offered himself up on Calvary for all mankind. All grace stemmed from this original sacrifice and its reenactments in space and time. A common Catholic custom was to offer sums of money for masses to be held for the souls of deceased loved ones—for their release from purgatory and ascent into heaven. Thus the sacral power of the priest extended over the dead.

Luther vituperated less about the parish priest but more about monks. One might think that Luther's paradox of thralldom and freedom applied to monks and nuns above all, since their raison d'etre was (usually) a *voluntary* thralldom in exchange for a spiritual freedom, imitating Christ's voluntary surrender of *his* freedom on the cross, only to later liberate himself and mankind through his resurrection. Luther's attack on monasticism is clearly crucial to the Reformation. The stumbling block was the ancient belief that celibate chastity was superior to marriage. Luther felt this to be one more claim of a *work*. Moreover, should monks and nuns remain frocked, they would be a constant reproach to Luther, who married a nun in 1525.

Since the very early church, virgins, male and female, had been held up as pure exemplars of Christian virtue who were remembered daily in the Canon of the Mass. Christ had spoken of this life when he invited the rich young man to give up all his possessions and follow him (Matt. 19:16–26). Additionally, he spoke of three types of eunuchs (Matt. 19:12), including those of the spirit who live for the kingdom of God. These texts are the Catholic justification of a poor and celibate clergy. Luther did esteem two medieval celibate saints—Francis and Bernard—but he denied that the precept of celibacy was widely feasible. Could Luther have been speaking from personal experience? It is certainly not impossible. In his own words, "Human frailty does not permit a man to live chastely, but only the strength of angels and the power of heaven." One scholar surmised, without evidence, that he may have struggled with masturbation when a monk. But the larger question was not Luther's fitness for celibacy, but celibacy's fitness for humankind. Let the monk or nun say "'I do not promise chastity'" or only "'so far as human frailty permits . . .'"—as faith and not as devilish "works."[8]

A year later Luther held up the "estate of marriage" as far superior to the cloistered state. Luther's attitude toward marriage is physiological as well as theological. He recognizes the imperiousness of sexual desire, as did St. Paul, and sees marriage as principally designed to allay passion. He reckons that only one in a thousand possess a celibate vocation. "There is therefore no comparison between a married woman who lives in faith and in the recognition of her estate, and a cloistered nun who lives in unbelief and in the presumptuousness of her ecclesiastical estate."[9]

A sequel to all of this is the Protestant doctrine of "callings"—Protestant because Calvin and Luther gave it high importance. Luther's priesthood of all believers means that there is no significant distinction between clergy and laity. Living *apart* from the world carries no worth in the eyes of God, says Luther. The Christian should live *in* the world. Differences of stations of life make no difference in heaven. Rulers are not more valuable in the sight of God than peasants, although in the world there are different callings and everyone has his rank and his place; nonetheless God equalizes them even here below. The sun shines on all alike. "Every Christian [is] an emperor and lord over all the masters of the world, not by reason of his person, in which he is a human being like the rest, but because of his faith in the Lord Christ."[10]

Concerning worldly ranks themselves, Luther follows the corporeal metaphors of St. Paul, who stressed that all are members of one body performing different but essential tasks for the good of the whole. Such was the favorite analogy of John of Salisbury, who developed it in his *Policraticus*. It operates in marked contrast to that of the modern competitive man driven by ambition. Luther advises: "Let everyone, in his manner of life, fill the office to which God has called him. Let him not exalt himself above others . . . Let everyone serve other men in love." He cites the Virgin Mary, whom Lutherans, unlike Calvinists, reverenced, whose humility shone in her daily performances of household and wifely tasks, and who did "not proclaim that she had become the Mother of God." The priesthood of all believers raises the ordinary human's view of himself. The ecclesiastical and political hierarchy weighs down on him far less. It displaces the medieval center of holiness from the pope onto the ordinary Christian. But Luther does not have a very dynamic view of human work, but rather a quite medieval and stationary one. His view of callings confirms the medieval, corporate view of the professions. Key to Luther's notion of *calling* (*beruf*) is the notion that man does not accomplish anything significant in his own work. Rather, he fulfills a role, wearing a mask that is God's instrument to accomplish his own work. To think otherwise would be to fall into the trap of the Catholic theology of works as opposed to the Lutheran theology of grace.[11]

Luther's juxtaposition of works over grace to the advantage of the latter underscores the complexity of Lutheran secularization. Preferring grace over works would seem to privilege a religious over a human viewpoint. But grace in Luther's hands is a scythe for cutting down what he sees as papal or clerical corruption and handing over the direction of spiritual affairs to local ministers (often of the people's choosing). All in all he is responsive to the congregation rather than to a distant, lofty, and authoritarian figure (the pope).

Luther may well have secularized the medieval world picture by disenchanting nature. The more impressive characteristic is his propensity to vilify, indeed diabolize, his enemies: the pope, the Catholic Church, Jews, Turks, sacramentalists (followers of Zwingli), Anabaptists, rebellious peasants, and others. These often get lumped together as when he writes, "The Pope, the Turks, the Jews, the common man and the Schismatic spirits do not know Him." Other times it is the pope who is singled out as the incarnation of the devil: "The pope is the true Antichrist and that all his doctrine, which he has extolled as the light of the world, is sheer devilish lying, stench and filth." Luther did not express any such hyperbolic hatred of the pope until he was excommunicated. It is quite possible that it was a projection or an expulsion of an inner sense of guilt brought on by that decision: if he weren't the devil, then the pope was. Luther redefines Christendom by damning the Turks, the Jews, and the pope who threaten its purity from the periphery. They all share a belief in justification by works, according to Luther. "It is the devil himself which directs peoples to their good work and not to Christ the Son of God." The church is the "devil's church," and "protectors of the church are just as blind and obdurate as the Jews were in their day." The Jews represent the law as opposed to the spirit; Lutheranism is St. Paul's Christianity directed against the Hebrews. The level of vituperation against the pope is equaled by that directed against the Jews. In his 1543 tract, "On the Jews and Their Lies" (which his translators reluctantly published), we find Luther railing against their prophets for paying more attention to glosses than the text (the presumed fault of the Papists as well), for their blasphemies against Jesus, for "poisoning our wells," and for being the "bloodthirsty bloodhounds and murderers of Christendom for more than four hundred years." As a "rejected and condemned people" their synagogues and schools deserved to be burned, and their houses razed. "Next to the devil, a Christian has no more bitter and galling foe than a Jew."[12]

Does this vilification have anything to do with secularization? I think it does in that it empties the church and the chosen people of their sacrality and diabolizes them. It marks a destabilization of the spirit—an anxiety about salvation that characterized a Calvin or a Loyola but also certain nuns, who were obsessed about the devil and feared his possession of their bodies and souls. Luther may have disenchanted nature of its medieval animism, but he diabolized human nature, curable only by grace and the Bible.

### iii

Luther was radical when dealing with the Catholic Church and its "works." But he was also intolerant of more radical readings of Scripture than his own, and even more so with social interpretations of the doctrines that were revolutionary. Some might say that this placed him in the position of a pope condemning dissent. In 1522 there was a rebellion of German knights by Franz von Sickingen, whom Luther warned against riling up "those who now lead the world so miserably astray." In 1525 it was the turn of the peasants who rose up in a massive insurrection against feudal land arrangements in Swabia and Thuringia. The title of his short tract "Against the Robbing and Thundering Hoards of Peasants" speaks for itself. Luther was unambiguous. Christianity does not condone rebellion. He roundly condemns attacks on property and goes even further to confirm the right of others to punish such attacks. Authority rests with God alone and with "rulers whom God has appointed."[13]

How do we interpret Luther's stout condemnation of social revolt at both extremes of the medieval hierarchy? Many writers, including Friedrich Engels, have scorned Luther's "reactionary" interventions. For them the Reformation is only the toppling of the medieval papal theocracy by the free German burghers. How could Luther have contradicted the revolutionary spirit so unambiguously once it appeared? The answer is that Luther, unlike Calvin, was interested in neither civil disobedience nor theocracy. The Lutheran church was to restrict itself to spiritual matters. It could accept a Catholic Charles V or a Lutheran elector of Saxony, Frederick "the Wise" or his nephew John Frederick (the latter two protected Luther). A separation of spheres, although never rigorous or absolute in Luther's mind, did secularize political office. Some historians see in this the beginning of the German cult of ecclesiastical obedience to secular rule. Luther's *Freedom of a Christian* and other pamphlets of 1520 had sown the seeds of social rebellion and sectarian extravagance. The Radical Reformation, as it is now called, turned on Luther for not going far enough, forcing him into the position of a firefighter fighting his own arson.[14]

To look at the Reformation's centrifugal effects we must first listen to the abuse that the radical reformers of Wittenberg, Zurich, and Munster heaped on Luther, just as Luther had heaped it on the pope. The eddies of rebellion did not stop at Worms or at the Wartburg but continued to multiply. Thomas Muentzer, radical Lutheran pastor at several churches and one of the leaders of the peasant rebellion of 1525, spoke already in 1521 of the reformers' being out of touch with their believers. In 1521 he infelicitously dubbed Luther a "single donkey fart doctor of theology." "The Lutheran clergy," he continued, "get their truth only from books and hearty flattery and pomp. But when God wants to write in their heart there is [*sic*] no people under the sun who are a greater enemy of the word of God than they." Luther is likened to a "fattened swine" and called a "doctor liar." Muentzer's antipapalism led to the opinion that "monks should be killed."[15]

Conrad Grebel, a patrician follower of Zwingli in Zurich, known for interrupting sermons he did not like, broke with that Swiss over the issue of iconoclasm leading the Zurich "brethren," in an Anabaptist movement of adult baptism. Grebel perceived that it was not just typical preachers who should be opposed, but "the anti-Papal preachers not in conformity with the divine word." It was no longer sufficient to be antipapal. Luther could not control these ripples of radicalization, even though he opposed them as firmly as the pope had him.[16]

Instead of listening to the doctors of Wittenberg, Grebel explains, one should taste the delicious experience of picking up the Bible oneself: "After we also took up Scripture and examined it on a great many issues, we became better informed . . . The evangelical [Luther's] preachers are silent about the divine word and mix it with the human." Grebel urges in this letter to Thomas Muentzer with forty signatures to abandon "the old customs of the Anti-Christ, such as the sacrament, the Mass, the sign, etcetera and to stick to the word alone." The radicals seemed to agree that an attack on images—including ones of the Virgin Mary—were in order to facilitate one's concentration on the word and spirit of God. Notorious acts of iconoclasm had been perpetrated in Wittenberg. Emphasis on revelation, both personal and biblical, invocation of the Holy Spirit to subvert structures, authorities, the written word, and conventions were the hallmarks of the Anabaptist revolution in Zurich, Holland, and eventually England. Such were the far-reaching interpretations of the experiences in Wittenberg and Munster. The fragmentation of Luther's religious message led to its sporadic reduction to a social and secular force that claimed even higher inspiration.[17]

The revolution of the spirit is perhaps the most unsettling of all Protestant theologies (and Catholic before them), because it could not be contained by Scripture, by priest (who must bow to the inspired believer), or by governor, whom they refused to recognize. Two customs in particular—adult baptism and polygamy—put Anabaptists at odds with every authority of Europe and was the principal cause of their persecution. Even though later sectarians condemned war and civil government, their vision of church-state relations are akin in some respects to Luther's: the Anabaptists accepted no dikes to Luther's freedom of a Christian, just as Luther accepted no walls around the papacy. Thus Hans Hergot, one of the radical reformers, speaks of the three ages of the Trinity when finally "God will humble all social estates, villages, castles, ecclesiastical foundations and cloisters." Such is how the hyperspiritualistic radicals used reformation as their vehicle of protest.[18]

### iv

In 1523, when he wrote *On Temporal Authority: To What Extent It Should Be Obeyed*, Luther was under the imperial ban. He was faced with two Christian traditions—one Pauline, which stated that "all authority comes from God" and therefore should be obeyed; the other medieval, which stipulated that Christians need not obey an immoral or unbelieving ruler. In addition Luther seems to have toyed with a third, antinomian position—that political authority should not be necessary at all, if only the world were entirely Christian: "If all the world were composed of real Christians, that is, true believers, there would be no need for or benefits from prince, king, lord, sword or law." Of course key suppositions here are the predominance of "real Christians" and "true believers." There can be little doubt that Luther identified such a community with no part of Europe, and certainly not with Anabaptist Europe, which took this doctrine to heart. "First fill the world with real Christians," Luther exhorted, "before you attempt to rule it in a Christian and evangelical manner." Without that, Christians need to be ruled "like wild animals so they don't devour others." Most men are evil and belong to the devil's kingdom and so need to be ruled by force. Complete regeneration would not happen. Real "Christians . . . are always a minority in the midst of non-Christians," he believed, echoing Augustine, his constant inspiration, and anticipating a more secular version of Robespierre that "virtue is always a minority on the earth."[19]

Luther abandons the notion of a completely regenerated kingdom of saints as indeed had Augustine before him. As such, he accepts, even to a greater degree than his medieval predecessors, the notion of a secular government separated from the kingdom of God on earth. The Christian submits to this secular rule of the sword for the sake of the world.[20]

But Luther goes much further in stressing obedience than had leading medieval theorists. The ruler is to be obeyed because he is appointed by God, even when he is evil. In fact Luther is credited with a theory of nonresistance, which he entertained perhaps in partial gratitude to the electors of Saxony such as Frederick III and John, who protected him and his church. The secular government is left to its crude and cruel activities such as waging war, burning, and plundering enemies' property. It does not live under a Christian order. Luther's dilemma was how to "satisfy God's kingdom inwardly and the kingdom of the world outwardly." The temporal is limited strictly to the earth and "has no authority over souls" but only over the "external dealings men have with one another." Expect not too much from such a government for "a wise prince is a rare bird."[21]

Luther radically separates the secular and the Christian, whereas medieval political philosophers in their "books of mirrors" had encouraged rulers to maintain Christian orthodoxy even by the sword. Luther is much closer to Machiavelli (whom he most likely never read) than to his scholastic forbears (whom he loathed). He would agree that the prince can and must do things that are not properly Christian. He must enjoy the full panoply of secular measures to achieve his ends—principally warfare. But if Luther does not see the ruler as an embodiment of or even constricted by Christianity, he *does* look to him for the maintenance of orthodoxy and the defense of Christendom from the Turks, and he even fantasizes the expulsion of Jews from the Western monarchies.

We can call this a political secularism because it vindicates a sphere removed from the inward spiritual man. In Lutheran terms the secular sphere is deprived of inner righteousness and freedom of the spirit. True, the Germans, Luther warns, must be cast out of their carnal existence—which can be identified as medieval externalism, the world of "works." But the Christian's weapons are not those of the world. One should cherish no illusions—the world of princes and popes (because the pope, for Luther, is a secular prince) does not listen to the word preached "incessantly to them." Only the sacraments are more powerful than they.[22]

A way of looking at Luther's secular political vision *in concreto* is to focus on his relationship to the Holy Roman Emperor Charles V. Having jettisoned the pope as the overlord of Christendom, does Luther acknowledge Charles in that role as the temporal arm of the papacy? This would be to establish a secular head of Christendom parallel to the secular heads of national churches in England and Scandinavia. Luther is consistently respectful toward the young prince Charles' "pious blood," after his accession in 1519. He writes to the bishop of Zwickau that Charles "is an excellent man [who] hopes to restore unity and peace" in divided Germany. He continues to urge unquestioning obedience to him: "Our regular governmental authority has been established by God and we are bound to obey." But he certainly sees the emperor's sovereignty as almost exclusively political and temporal: "the earthly kingdom that is man's bodies and goods. There his office ends." He declines to attribute to him any role in the suppression of heresy (which would be Lutheranism!). Thus Luther anticipates Locke's stripping the state of any spiritual function.[23]

Luther saw Charles, Catholic prince that he was, as his main hope in dealing with Rome and in getting a church council called, preferably in Germany, without the presidency of the pope. It is astounding that the religious differences between Charles and Luther did not cause them more problems. Luther stakes out his doctrine of nonresistance unambiguously:

> Even if . . . His Imperial Majesty acts unjustly and operates contrary to his duty and oath, this does not nullify the authority of the Imperial government, nor does it nullify obedience on the part of the emperor's subjects . . . Even if an emperor or sovereign acts contrary to all of God's commandments he still remains emperor and sovereign."[24]

The medieval deposition and interdict are gone. As long as the empire and the electors consider His Majesty to be emperor, and do not remove him from office, he is to be obeyed.

Ultimately princes stand or fall according to God's sovereignty, not papal suzerainty or that of one's subjects: "With a single word or nod He shatters the spirits of the greatest warriors . . . [He] takes away the spirit of princes and is terrible to the Kings of the Earth." For "it is ungodly and useless to place confidence in fortifications . . . ramparts and guns." Luther felt vividly that God, not the emperor, was the supreme authority.[25]

This conviction is well illustrated in the diatribe that he penned against Duke Henry of Braunschweigner: "Unless the Emperor is submitted to God, and obeys the fourth commandment, coronation is pointless." Luther brandishes this rod of divine deposition in the face of all earthly potentates who travesty Christianity in neglecting to recognize Christ as their spiritual ruler. Peasants also adulterate the Gospels, he might also have thought, by giving them a predominantly social-economic interpretation. Both are secular interpretations. With one sleight of hand, Luther empties the indelible sacred power conferred by medieval coronation. At no time was this secular vision of political power as evident as during the apocalyptic siege of Vienna by the Turks in 1529, when Luther counseled Christians to cooperate with Charles, not as a leader of Christendom waging a crusade, but only as a "secular ruler." The empire was being demystified, as the papacy had already been. All that was left was man and God, who was no longer immanent, but indeed "hidden." By April 1539, in the middle of the Schmalkaldic wars between Lutheran princes on one side and the papacy and the emperor on the other, Luther did modify his position on civil obedience and justified war against the pope as the Antichrist and the emperor as *miles papae,* or the "pope's soldier." This was perhaps not so much resistance to one's ruler but self-defense against foreign aggression—that is, Charles's last and unsuccessful attempt to stamp out Lutheranism in Germany.[26]

Luther's reformation had inherently secular tendencies that inspired vehement dissidence at home and abroad. One instance was his controversy with Erasmus in 1525 over free will. Another was with Zwingli about the real presence of Christ in the host in the Marburg Colloquy of 1529. While these disagreements produced skepticism about religious truth itself, the full consequences were not felt until the next generation. But the disputes of the first-generation thinkers among themselves created sufficient doubt in the minds of the second-generation thinkers as to whether *any* part of the Christian tradition could be proven.

The historian Gerald Strauss, who holds the view of the inefficacy of the Reformation at the popular level, cites a memorandum of a Saxon jurist, Melchior von Osse, to Duke August of Saxony in 1555. Incessant disputes on the fine points of theology and administration were causing "lay people and common folk to doubt the very articles of the faith and to hold the preachers, indeed the entire religion, in contempt."[27]

The sociology of religion as practiced by Gabriel Le Bras, Keith Thomas, Carlo Ginzburg, and Jean Delumeau has emphasized the extent of popular unbelief and resistance to reform. Thus instead of the mythical "mount of piety" around 1300 from which all Catholic modernity descends, historians like Keith Thomas posit a vast grassroots obliviousness to religion, a predilection for superstition and sometimes even atheism, or a spontaneous, unreformed belief.[28]

However true this may be (the sources of popular culture may never be adequate to tell), the Reformation ran into opposition not only from emperor and pope but from within its own sectarian ranks and congregations. The logic of the phenomenon is exceedingly simple: if reformer A contradicts reformer B, how do I know which is right? Multiply the As and Bs sevenfold, and the question arises, Is there any truth in it at all?

Richard Popkin, in his *History of Skepticism: From Savonarola to Bayle,* devotes one chapter to the role of the Reformation in the rise of seventeenth-century skepticism. Skepticism existed on the popular level with a shrug of the shoulders and an inarticulate refusal to believe what seemed to contradict common sense. At the learned level, of course, it was much more elaborate. Erasmus and Luther debated over free will, Erasmus maintaining its necessary connection to merit and sin and Luther decrying it as work or righteousness usurping the role of divine grace. Once the stakes of Scripture versus tradition were planted by John Eck, the papal defender against Martin Luther in Leipzig in 1519, it was incumbent upon their audience and followers to accept one or the other. The interpretation of Scripture was too fraught with ambivalences to be accepted as self-evident. What is meant, for instance, by "Thou art Peter, and upon this rock" (Matt. 16:18); "If thy foot is an occasion of sin to thee cut if off" (Mark 9:44); or "Take and eat this is my body" (Matt. 26:26)? Councils enjoined interpretations. Protestants believed either that one was to be taught by authoritative reformers such as Luther and Calvin or that the individual, on his own, could make decisions concerning Scripture. But the inconsistency is that Luther, in practice, considered those who differed from him to be "schismatics."[29]

An extraneous body of ideas appeared in the 1562 translation of works (known earlier) of the third-century Greek empiricist Sextus Empiricus, who introduced rigorous academic doubt into philosophy and religion from which there was no apparent exit. For Sextus one had to find a principle of certitude, but doing so simply begged the question of *its* principle of certitude, thus regressing ad infinitum.

Aristotle and Thomas had been under attack almost from the latter's last breath. Erasmus, Luther, Pico della Mirandola, and Lefèvre d'Etaples had chipped away at their reputations. Montaigne, Bacon, Locke, Descartes, and many others would hand over a mutilated effigy of the Stagirite to the Enlightenment.[30] The greatest philosopher of the era of the French religious wars, Michel de Montaigne, plumbed the deepest skepticism of the age. Nurtured by a Protestant mother, whose ancestry traced backed to Portuguese "new converts," and a Catholic father, Montaigne was educated in distinguished schools in Bordeaux and Toulouse, where humanism and Protestantism had spread. In Paris he brushed against Jesuits, who appeared shortly before Luther's death to meet the new crisis of belief. Furthermore Montaigne took a keen interest in savages without *any* Christian background, whom he heard about from the travel reports brought back from the New World. This mix made Montaigne a philosophical Pyrrhonist, a cultural relativist, and a counterreformation fideist. Pyrrhonism is the philosophy that nothing can be known with certainty, least of all perhaps the Aristotelian truths that the church had made its own in the High Middle Ages as a substructure to Christian doctrine. It dovetailed Ockhamistic nominalism, which had attracted Luther as a theology student. Fideism can be quite modern and quite epistemologically corrosive. The Thomistic, Dantesque, and Jesuit syntheses saw classical reason and scholarship neatly corroborating faith. Fideism had the air of "pleading too much" while supporting too little. "True religion," fideists said, "can only be based on faith; . . . any human foundation for religion is too weak to support divine knowledge." This ostensibly orthodox counterreformation maneuver had the effect of emancipating philosophy from the tutelage of faith and setting it loose to ruminate upon itself. It also had the secularist effect of removing religion from a discourse of reason, excluding it as extrarational and therefore not to be discussed in raison d'état or reason tout court. Counterreformation fideism and Pyrrhonism secularized the public sphere.[31]

Descartes inherited the Pyrrhonist legacy. He was not a fideist, but he did not think that religion (or politics) could be made part of the new philosophy, because it was the product of custom and opinion. It had to be secluded therefore—put aside (put asleep?) in what Pierre Chaunu has aptly called the Cartesian parenthesis, a parenthesis made all the more tempting by the fate of Galileo. This leaves the next installment of the story to Locke—a Calvinist by upbringing, a skeptic by experience, and a firm believer in a secular solution of the Reformation legacy.[32]

## V

Luther's secular thrust is paradoxical because it is clothed in religion and is visible mostly in its effects and consequences than in its initial thrust. But even that first stirring can be considered secular in its critique and denial of the spiritual purity of the church. In repudiating Catholic claims to suzerainty over Christendom, Luther contributed to its eclipse of overall spiritual leadership—one, it is true, that had faltered in recent decades. Hence a greater secular space emerged.

It is part of the process of the Reformation to use a higher divine authority to reduce a human authority, judging the temporal authorities by an appeal to God. Luther objected to this practice in the radical reformers, who appealed over his head to the Holy Spirit. But he had begun the whole approach at Worms when he refused to budge unless corrected by Scripture and evident reason as opposed to the authority of the church. Such appeals became endemic—indeed the very essence of the Reformation—as one preacher challenged another with *his* scriptural passage or inspiration moving from Zurich to Basel, to Strassburg, to Antwerp, to London, to Boston, to Providence, to New Haven.

In this chapter we have been interested in the contribution of the Lutheran reformation to modern secularism. We have tried to point out the paradoxical nature of this contribution: how an emphasis on religious phenomena such as faith, Scripture, and grace—so rich for so many adherents who found deep spirituality in Luther's message—had a secular effect of undercutting traditional Christendom and identifying it with its popular enemies, the Turks and the Jews. One of the most salient characteristics is Luther's displacement of religious authority and sacredness from the clergy to the laity expressed in his doctrine of the "priesthood of all believers." Another is the relationship between subject and sovereign, the ultimate recourse for Luther being the sovereign's subordination and subjection to God. This is in continuity with the entire Christian era, the difference being that it is not mediated by the church. The later loss of all subordination to the church and God is arguably a precondition of modern totalitarianism, the essence of which is the ruler's (or people's) sense that he is responsible to no one but himself. Third is Luther's deference to secular political authority. His attitude toward Charles V, as a desacralized ruler, hitherto crowned by the pope, speaks for itself. A fourth finds Luther at bay vis-à-vis the multiple, often violent, interpretations of the Protestant Reformation over which he had no control and little sympathy. This cell-splitting phenomenon is often seen as affording the Christian believer the maximum pluralism and religious liberty possible. Unfortunately, its adverse side effects were the despair of people such as Montaigne and Descartes, who gave no rational credence to any existing creed. Rather, they privatized or parenthesized them. Within a century, if religion were to occupy a place in the public sphere, it would most likely not be its center. The Jesuits certainly succeeded in temporarily integrating papal loyalty and humanist education in their several hundred colleges. But there grew up in the seventeenth century the recourse of reasoning without reference to Christianity (Pascal was certainly countercultural), of seeking refuge in scientific academies away from religious enthusiasm. The effects of secularization were often more the result of elliptical displacements than confrontational oppositions. Luther certainly fulminated against the pope most vilely, but his secularism lay primarily in the transferal of sacredness from the clergy to the laity, from the pope to the prince, and in the proliferation of variant evangelia among his followers such that John Locke in the next century would despair of establishing orthodoxy itself.

# Part III
## Introduction:
## Autonomy in the Enlightenment

Early European modernity strongly emphasized the individual as the source of authority and truth. Descartes advocated universal doubt. Locke said every man was orthodox to himself. He was to obey only himself, said Rousseau. Morality, said Kant, was making up a rule oneself that everyone should follow.

This, of course, seriously undercut the authority of churches, which traditionally prescribed duties to their followers. Locke and Rousseau, however, felt that the individual could reach mundane truths on his own, but they questioned Christian metaphysics and mysteries that went beyond these truths. Kant went ever further, denying that anyone could know by "pure reason" the existence of God, the freedom of the will, and the immortality of the soul. But he conceded that it was good "for practical reason" to assume that they were true. Destutt de Tracy, in his turn, pushed empiricism to its limit, denying that man knew anything he could not actually "sense," which he equated with thinking.

All these men believed that the individual had to examine his reason introspectively and discover what he could know and what he could not know. The irony of the age of reason is that the range of man's reason was greatly diminished. If man could not know the answers to the eternal questions, which they deemed he could not with certainty, he must focus his efforts on more limited fields: politics, economics, and peace. Here the individual could play a decisive role as a party in the social contract, a participant in elections, a holder of natural rights, a consumer, and a holder of property. Nothing stood between the individual and the state either. The French Revolution wiped away all those medieval corporations, all those "intermediary powers," buffers between the individual and Leviathan. In all these cases a new secularity had been forged: Locke, by denying a Christian orthodoxy; Rousseau, by separating Christianity from the state, and Kant, by removing Christianity's metaphysical base.

# Chapter Seven
# Locke: Toleration, Infallibility, and the Secular State

John Locke is *the* philosopher of the Anglo-Saxons. He rebutted extravagant Cartesianism with his commonsense, empirical psychology. He championed consensual parliamentary politics and economic individualism. Against a Stuart Catholic Restoration, he spoke out for religious toleration. These ideas were exported to the American colonies, where they enjoyed wide circulation and approbation. They are, moreover, in some way the root of our present cult of "diversity." Here we investigate the source of Lockean toleration and relate it to the whirl of religious controversy in the English civil wars (1640–1660) and the Stuart Restoration (1660–1689).

Locke is incomprehensible if seen apart from the Puritan revolution that raged during his childhood. But he stands out because he embodied a secular conclusion of the Protestant Reformation rather than the sectarian tendencies it spawned.

We have argued in the preceding chapter that Luther's reformation was centrifugal, that instead of giving rise to one reformed church it spawned dozens. This was the price of recognizing the freedom of consciences and biblical interpretations. As a young scholar at Oxford in the 1660s, Locke was acutely aware of these problems and reacted quite differently to them than he would when confronted after 1679 by Stuart Catholicism.

Each of the Protestant reformers believed that he had the true view of Christianity, for each had gone back to its inerrant source, Scripture, bypassing the pope, the church, and tradition.

In England the Tudor monarchy had contained Protestantism to enact minimal reforms for a century. But during the subsequent Stuart reign (1603–1649), High Church orthodoxy (Anglicanism) and political absolutism clashed with Puritanism and parliamentarism. Civil war broke out in 1640, leaving its mark on the young Locke. Presbyterians, Independents,

Levelers, Diggers, Quakers, Ranters, and others all had visions of what the ideal commonwealth and church should be.

Locke was born in 1632 at Wrington near Bristol. His father was a country attorney of Puritan sympathies who sent his son on scholarship to Westminster School in London. The young man matriculated at Christ College, Oxford, in 1652. After two decades of revolution and Cromwellian dictatorship, the Stuart dynasty was finally restored. As a lecturer at Oxford in the 1660s, Locke pondered the question of how civil peace could be ensured and penned two rather scornful pamphlets on ceremonies and rituals of these diverse "dissenters." He looked to the restored monarchy under Charles II to quell this religious effervescence. But when Charles's brother, James, a Roman Catholic, emerged as Charles's successor, Locke went into exile as a Protestant resister and began writing prolifically for toleration, concluding that no one religion could be proved the true one and that "everyone is orthodox to himself." King James alarmed Protestants, who feared that Roman Catholicism was in the ascendancy, orchestrated by Louis XIV's absolutism in France and his revocation of Protestant liberties in 1685. English liberties could only suffer from these eventualities. Toleration, which Locke had once thought should be curbed, seemed far preferable in 1689.

Although France had taken decisive steps toward religious uniformity, many of its philosophers, especially those known as *libertines,* and skeptics, such as Michel Montaigne, Pierre Charron, and Pierre Bayle, were edging away from orthodoxy. The Protestant Reformation had led many to doubt not only religion but truth itself. Descartes began with skepticism—in fact, "universal doubt"—before he discovered the certitude of personal existence in the *cogito* ("I think, therefore I am"). But Descartes avoided questioning religion, excepting that of his country, because he did not think that either religion or politics was subject to philosophical proofs.

Locke wrote his *Essay Concerning Human Understanding* (1690) with similar preoccupations about what could be proven true, but he used empiricism rather than Cartesian rationalism to ruminate about doubt. In his works he subjected religion and politics to rational analysis—a major difference from Descartes'.

Locke's secularism consists of a certain despair about resolving differences among Christians, and he concludes that "everyone is orthodox to himself." When he writes about Christianity, he omits key questions such as original sin, the divinity of Christ, or the redemption. He seeks a lower common denominator in religion—in the hope that it will generate agreement among the two percent of the population considered "gentlemen."

Instead, it led to a further reduction of beliefs—those that Locke considered indubitable were subsequently doubted. Locke served as an intermediary between the Protestant Reformation, in which he was cradled, and the French Enlightenment, which he greatly inspired. His reduction of the role and content of religion led to its marginalization, or, indeed, was itself a kind of marginalization.[1]

### i

*Two Tracts on Government,* written in 1660, was known only to scholars before its publication in 1967, when its editor referred to the philosopher-author as "the conservative Locke." Two additional tracts of the same decade, on infallibility and toleration, written while Locke was at Christ Church, Oxford, reveal him as censorious of the religious legacy of the civil wars, as Hobbes had been of its politics. They sketched Locke's position on religious pluralism and governmental uniformity and led him ultimately to make some of the most remarkable secular pronouncements in the wake of the Glorious Revolution.

In 1660, Locke was accurately described as a latitudinarian, that is, one who wished to open the door of the Anglican church to a wide number of worshippers by requiring belief in only a small number of dogmas and ceremonies. Nowhere does he prescribe acceptance of the Thirty-Nine Articles or the Book of Common Prayer, staples of orthodoxy in the sixteenth century. Locke was not a dissenter, because he did not belong to any of England's major sects—Congregationalists (Independents), Presbyterians, Baptists, Quakers, and Catholics—that were excluded from the toleration provided by the Restoration's Clarendon Code (1661–1665). In fact he was quite scornful of these sects.

In the *First Tract on Government* (1660), Locke admitted that he was a "professed . . . enemy to the scribbling of this age," because the pen is as guilty as the sword of the "Furies, War, Cruelty, Rapine, Confusion etc, which have so wearied and wasted this poor nation." In characteristic seventeenth-century fashion, he italicized key words to dramatize the Puritan and sectarian controversies from which England was emerging. The times, he wrote, are fraught with "*disputes,*" "*disquiets,*" "*troubles,*" "*discontents,*" "*doubts,*" "*private judgment,*" "*oppression,*" "*disorder,*" "*tyranny,*" "*anarchy,*" "*general bondage,*" "*giddy folly,*" and "*contention, censure and persecution.*" In the *Second Tract on Government* written in Latin, probably before 1662, Locke is no less peevish. The civil wars, he argued, were sparked by "bitter party quarrels" and religious "enthusiasms," both of which he found very dangerous.[2]

The question Locke tried to answer in the longer *First Tract* was "whether the Civil Magistrate may lawfully impose and determine the use of indifferent things—in reference to Religious Worship." We know by his "Essay on Infallibility," written in this period but published only in 1977, that Locke believed that a magistrate was infallible in what he called "indifferent things"—the peculiarities and variations of religious customs and ceremonies, such as the Quaker practice of wearing hats in church and the Baptist practice of adult baptism.[3]

Locke's correspondence reveals that he was somewhat intolerant of ritual or "ceremonial" differences that sectarian controversy produced. He thought that the government should step in and make peace. In a letter to his father, he cited a Quaker woman, who, with several coreligionists, was summoned to Parliament, where she made "a continued humming noise longer then the reach of an ordinari [*sic*] breath . . . [while] another sung 'holy, holy, holy'." "I am weary of the Quakers," Locke concluded, "and returne to my friends with you."[4]

As an ambassador's secretary in Clèves in 1665/66, Locke observed a pluralistic society where Lutherans took unconsecrated "wafers" [*sic*]. This practice, he grouched after an exhausting argument, left him with "nothing but some rubbish of divinity, as useless and incoherent as the ruins the Greeks left behind them." Surprisingly, he reported, "I have not met with any so good natured people or so civil as the Catholic priests." But in a letter to Henry Stubbe, he already staked out his opposition to toleration of Catholics, a position he did not change.[5]

Locke might even have been harsher on Protestants than on the Catholic priests, because he faulted the former for unleashing the religious "enthusiasm of the Civil Wars" that were fought partially over ritual. These "indifferent" yet inflammatory matters, such as dress and posture at divine liturgy—items not prescribed by the law of nature or by divine law—should be left to the magistrate, who should have the final "infallible word" for the sake of peace. Locke's magistrate has been called an "umpire," who would establish rules by which sects could coexist peaceably, and not create exclusive rules and practices.[6]

Does endowing the magistrate with authority make Locke a supporter of religious uniformity? Does he suppress, as some accused him, "Christian liberty" by requiring sects to abandon their religious idiosyncrasies?

Locke was definitely apprehensive of complete religious liberty: Early on he wrote, "Grant the people once free and unlimited in the exercise of their religion and where will they stop, where will they themselves bound it, and will it not be religion to destroy all that are not of their profession?"[7]

He even attributed the source of seventeenth-century war, particularly the Thirty Years' War, to sectarianism wearing the "visor of religion":

> The cunning and malice of men [have] taken occasion to pervert the doctrine of peace and charity into a perpetual foundation of war and contention, all those flames that have made such havoc and desolation in Europe, and have not quenched but with the blood of so many millions, have been at first kindled with coals from the altar.[8]

But far from blaming Christianity for this destruction, as would Voltaire, Locke blamed only its prevarication. True Christian liberty was not the cause of these wars, but their casualty. Locke was as interested as anyone in seeing a true Christianity flourish. He too wished to protect its inner spirit, which was not possible if sects were given complete freedom. In 1660 he appeared to be more a defender of English conformity than nonconformity. He failed to see any incompatibility between an exterior conformity and an inward Christian freedom. "God may be worshipped in spirit and in truth as well where the indifferent circumstances are limited as where they are free," he wrote. "A gracious heart may pray as fervently in the ancient form of the Church as [with] the extemporary form of the minister."[9]

Locke's uniformity does not even stop with curbing external practices. He was wary of the term "freedom of conscience" and thought it was not the magistrate's rule that affected it but only the "indifferent" external things of ceremony.[10]

Locke's secularism, at this point, is Erastian and reductionist. Government should co-opt a sphere of religious activity of "indifferent things," he wrote, because one could "comb the Gospel to no purpose, for a single common standard of propriety" regarding those "indifferent things, even those regarding divine worship, [which therefore] must be subjugated to governmental power." In controlling religion's external manifestations, the state could claim infallibility in matters formerly reserved to a church, and religion would be partially secularized.[11]

## ii

In 1667 Locke authored two more essays, which remained unpublished during his lifetime: "An Essay on Toleration" (not to be confused with his 1689 *Epistola de Tolerantia*, translated as *A Letter Concerning Toleration*), and the Latin *Essay on Infallibility*. In these essays, Locke has moved closer to nonconformity and shows that he is less eager to have the government regulate religious matters and more willing to let the individual exercise his freedom of worship. No longer does he regard ceremonies as the special jurisdiction of an infallible magistrate, nor does he think that doctrine or morals should be removed from the magistrate's purview. Indeed, he dismisses "mysteries" such as the doctrine of transubstantiation and the Trinity, which, he says, cannot be interpreted because they cannot be understood. Foreshadowing later controversies surrounding his *Reasonableness of Christianity* (1695), he argues that revelation can be understood without an interpreter. Overriding all his former criticism of disruptive sectarianism, he now concludes that the evangelists were sufficient interpreters; there is no need of an infallible one, for Scripture is plain to the understanding.[12]

To a Catholic, the multiplicity of interpretations is sure evidence of the need of a single inerrant interpreter. But Locke turns the argument around and says that the multiplicity of interpretations proves the nonexistence of such an infallible teacher! Inerrancy rests in Scripture alone, Locke says, falling back on Luther's *sola scriptura*. But then, one may ask, if the meaning of Scripture is so evident, why are there so many interpretations? At this juncture Locke clinched his new view of depriving the magistrate of any pastoral role: "The magistrate hath nothing to doe with the good of men's souls. . . ."[13]

Locke's resulting state is almost deconfessionalized and secular. It does not enjoy divine right (a theme later developed in his *First Treatise of Government,* 1689/90). It does not even seem to enjoy a fragment of infallibility (in "indifferent things"). Locke is clearly out of step with the divine right being claimed by both the Stuarts and Louis XIV.[14]

The individual is the main beneficiary of Locke's reworking of his youthful political theory. Locke seems now to transfer to the individual the infallibility he earlier relegated to the ruler. Thus, he now revisits the details of religious worship that riled his sense of public order: "kneeling or sitting" while taking the sacrament; "wearing a cope or a surplice in the church"; being baptized in adulthood or infancy; and observing the Sabbath on "the Friday with a Mahometan, or the Saturday with the Jew, or the Sunday with the Christian, "with "the various and pompous ceremonies of the papists, or in the plainer way of the Calvinists." He sees "nothing in any of these, if they be done sincerely and out of conscience, that can of itself make me either the worse subject . . . or worse neighbor."[15]

In other words, the state should not regulate these matters, precisely because they are unimportant. Even dubious moral practices, such as polygamy and divorce, merit toleration since they are indifferent, he says;

they do not affect the public order. The magistrate can act as an umpire, but not as an infallible one.

> For, not being made infallible in reference to others by being made a governor over them, he shall hereafter be accountable to God for his actions as a man, according as they are suited to his own conscience and persuasion; but [he] shall be accountable [to man] for his laws and administration as a magistrate, according as they are intended to the good preservation and quiet of all his subjects in this world as much as possible.[16]

This "umpire," while not infallible, has a vast, hypertrophied competence that can eclipse even divine laws. "The good of the commonwealth," Locke writes, "is the standard of all human laws, when it seems to limit and alter the obligation of the laws of God and change the nature of vice and virtue."[17]

Whether religious or not, strange behavior is no longer qualified as "indifferent" when it "appear[s] dangerous to the magistrate." Catholics cannot be tolerated because they do not advocate tolerance—and because they take an oath to the pope as "the sole and infallible interpreter of the Holy Bible" and head of a foreign and hostile state, which could negate obedience to the king.[18]

Catholicism, in short, was "subversive of civic authority" and "absolutely destructive to society." Anti-Catholicism, the foremost English phobia fed by Bloody Queen Mary, Guy Fawkes, and soon, Titus Oates, targeted less than 1 percent of the population. Reconfirming Tudor repudiation of papal infallibility, Locke attributed some of its erstwhile competence to the Erastian state (not without inconsistencies), which Hobbes had recently designated "Leviathan." Both men subscribed to the fundamental law of the Tudor Reformation, namely, the Act of Supremacy of the state over the church, which papal authority necessarily challenged. Locke supported this Erastianism while depriving the state of any doctrinal or religious authority, properly speaking. In short, it was secular.[19]

Then where is infallibility in religious matters if it is neither in the pope nor in the magistrates? A kind of makeshift, ad hoc infallibility can be found in the "shepherds" (pastors), not so much because they are inerrant but because the flocks can do no wrong in following them. This could be called "a representational infallibility." It is "directive not definitive," founded on pragmatic grounds (the necessity of obeying *someone*). The only real infallibility lies in the Bible itself: "The most certain interpreter of Scripture is Scripture itself, and it alone is infallible."[20]

How consistent are Locke's arguments in these four essays? In the 1660 *Tracts of Government,* governmental infallibility exists in "indifferent matters." In the 1667 essays he transfers this infallibility to the churches, which can speculate on dogmas, such as the Trinity, which are above reason. Petrine infallibility, found in Matthew 16:18–20, is, surprisingly, not mentioned. The infallibility that Locke alludes to is not, it seems, absolute, or pontifical, but pragmatic and secular, invoked to achieve peace among the sects. The magistrate, the umpire, will tolerate as much as possible, for Locke recognizes that persecution is fruitless because it only strengthens the convictions of the persecuted. Toleration, then, is here acknowledged on the grounds of raison d'état, or Lockean Machiavellianism.[21]

It is not difficult to fault Locke's solution to the problem of religious pluralism. To start, one of the reasons for religious strife was individual interpretation of Scripture. Locke elliptically recognizes this when he says, "Anybody may attach a new meaning to the words [of civil and religious laws] to suit his own taste." The Bible does not contain or endorse the concept *sola scriptura.* Luther and Calvin expected their followers to interpret the Bible much as they themselves did. Locke recognized the dangerous "fancy of every interpreter" but implicitly embraced individualism as preferable to papal authority.[22]

The establishment of the supremacy of the state is a crucial act of Reformation secularization. Secular control of church affairs was recommended as far back as Marsilius of Padua's *Defensor Pacis* (1324). Even the Catholic states of Spain and France acquired through concordats with Rome the privilege of nominating their bishops. Locke's state had the right of making the last call, even though it might not be absolutely correct. What he valued above all was the peaceable individual who would interpret the Bible on his own, free of all interference from church and state.

Locke's quest for these benefits of peace and tolerance among church, state, and individual coincided with the dawn of an era of stability in early modern Europe. The Thirty Years' War had ended in 1648; the civil tumult in France, known as the "Fronde," was quelled in 1653; and the long and vigorous reign of Louis XIV (1661–1715) had begun. Europe, according to one eminent historian, Theodore Rabb, had achieved "stability," the backdrop for a confident enlightenment.[23]

### iii

In 1678 Titus Oates discovered the Popish Plot, by which Jesuits allegedly sought to assassinate Charles II to hasten the accession of his Catholic brother, James. This discovery led to the Exclusion Crisis of 1679–1681, an attempt to bar James from the throne. In the ensuing crisis, many Catholics were killed. Locke was the adviser-confidant of Lord Ashley, third earl of Shaftesbury, who was the leader of the Protestant and parliamentary cause against James and the leader of the fledgling Whig party. Shaftesbury and Locke had subscribed to the two Test Acts of 1673 and 1678, which required repudiation of the Catholic doctrine of transubstantiation. Shaftesbury stood for Protestantism, representative institutions, and the supremacy of Parliament, and Locke derived his whiggism from him. Their common dislike of Catholicism put them on a collision course with James. "Toleration" was their antidote to religious and political absolutism, "a defense against popery, arbitrary power and tyranny." But Shaftesbury's support of exclusion led him to the Tower, from which he soon escaped to flee to Holland, where Locke joined him in September 1683. It is most likely that during the Exclusion Crisis in the early 1680s, Locke began writing his *Two Treatises of Civil Government*—a response to Stuart absolutism as well as to Robert Filmer's *Patriarcha or the Natural Power of Kings* (1680), which argued in favor of primogeniture, or patriarchal succession of kings from Adam on down. In his second treatise, Locke argued that individual consent, rather than divine right, was a prerequisite to establishing any government.[24]

Louis XIV's absolutism, rather than English parliamentarism, seemed to be the wave of the future. Charles died in 1685, and James *did* succeed him to the throne. James Scott, Duke of Monmouth, who aspired to a Protestant throne himself, raised an army but was defeated at Sedgemoor and executed by King James. Meanwhile, by the king's direct order, Locke was deprived of his "studentship" at Oxford, a sinecure that had brought him income and few duties since the 1650s. Apparently, this was just what Locke needed! He now went into a period of intense literary creativity. In addition to continuing to work on the *Two Treatises of Government,* he wrote drafts of his *Essay Concerning Human Understanding* and his *Epistola de Tolerantia,* works that he would have to explain and defend for the rest of the century. The last was translated and published in English by the Socinian (anti-Trinitarian) William Popple, without Locke's knowledge, in 1689. In short, Locke was writing, publishing, being translated, and becoming famous. As his best biographer put it, "In the later months of 1689 Locke had no fewer than three books in the press, and what is more, these three were the most important he ever wrote."[25]

While in Holland, Locke made lasting friendships with Philip van Limborch and Jean Le Clerc, the editor of one of Europe's first learned periodicals, the *Bibliothèque Universelle,* which published a hundred page *abrégé* of Locke's *Essay Concerning Human Understanding* in the January–March issue of 1688, before it appeared in London in 1690.[26]

But tolerance was not unlimited, even in Holland, although it enjoyed the reputation of being a refuge for victims of religious persecution, such as the Marian exiles in the 1500s and the French Huguenots in the late 1680s. Limborch himself was working on the horror-filled *Historia Inquisitionis,* about which he corresponded with Locke frequently over several years. In March 1689, Locke expressed to Limborch that he hoped to see a broadening of tolerance in the Church of England. Limborch agreed that a "Catholic and Christian Church," that is, one favoring freedom and diversity of doctrine and organization, was desirable. His praise of Locke's now finished *Epistola de Tolerantia* was unbounded: "I have never read anything on this subject which appealed to me so strongly." On the same subject, with another continental, Nicholas Toinard, Locke corresponded about a *harmonia* of the gospels—a work Toinard had worked on for years and that interested Locke because of its irenic and latitudinarian implications. These men—Locke, Le Clerc, Limborch, and Thoynard—exchanged books on the most audacious theology back and forth across the Channel right into the eighteenth century.[27]

This interest in religion among men of science and letters was hardly unique to Locke's circle. Robert Boyle, Locke's "chief scientific mentor," was an "earnest Christian," who reconciled the new science with Anglican orthodoxy. Isaac Newton, Locke's good friend and occasional correspondent, was keen on reinterpreting the Pentateuch, especially Genesis, where he found the origin of monotheism. For Locke's friends and correspondents, science was not the problem, but rather history, opinion, and politics.[28]

Locke sailed back to England in February 1689 aboard the same ship that carried Mary, James II's daughter, the new queen of England. In London, he was greeted by the Act of Toleration, which had been drawn up independently of his *Epistola* and granted by the new sovereigns.[29]

Was Locke satisfied by this extension of religious liberty to all but Roman Catholics? His correspondence reveals little about the Glorious Revolution (which brought William of Orange and Mary to power), yet one can surmise, in the light of his and Shaftesbury's role in the succession crisis, and his later close association with the Duke of Monmouth's family, that he would have thought exclusion of Catholics even more urgent then than ever before. Indeed the Glorious Revolution was accompanied by a rash of anti-Catholic legislation in Maryland, Massachusetts, and other colonies.

Locke left no doubts about his stand on the revolution. He wrote of the looming conflict in Ireland between James II and Schomberg, William and Mary's general: "The campaign in Ireland is going forward well enough, and if God wills there can be little doubt of its success." The defeat of James II in the Battle of the Boyne in 1690 settled Protestant domination in Ireland and England for the foreseeable future.[30]

The question Locke asked in his *Letter Concerning Toleration* was, How should the Commonwealth treat the church? For Locke the "chief Characteristical Mark of the True Church" was toleration, not doctrinal uniformity. And if the church were tolerant, it must necessarily accept all the variations in dogma that had come into existence among Protestants. *Caritas* or love, not doctrinal feuds, was the sign of Christianity. The heart, not the intellect or a required belief, which would only produce strife, martyrs, or hypocrites, would unite Christians. With this, Locke sanctioned doctrinal fragmentation and the total relinquishment of a state-imposed orthodoxy, which he thought was incompatible with individual freedom.[31]

The case for orthodoxy of the English church was already greatly compromised by its sovereigns. "Our modern *English* History affords us fresh Examples, in the Reigns of *Henry* the *8th, Edward* the *6th, Mary* and *Elizabeth,* how easily and smoothly the Clergy changed their Decrees, their Articles of Faith, their Form of Worship, everything according to the inclination of those Kings and Queens . . . [which] no man in his Wits [can] obey."[32]

The Tudors' intolerance was little in keeping with the spirit of the Gospels, Locke felt, but the clergy were too supine to resist their sovereigns' doctrinal demands. The combination undermined the credibility of the Anglican church. Nor were Lutheran inwardness and Calvinist rectitude well attained by doctrinal formulas or tests.[33]

Locke's main concern, however, was not with credibility or uniformity—he had given that up if he ever had it—but with the individual's right to consent, which was the foundation of his political and social theory, as laid forth in the *Second Treatise of Government* (1689/90), *Some Thoughts Concerning Education* (1693), as well as *A Letter Concerning Toleration* (1689) and its second, third, and fourth sequels. No government has the right to command an individual's obedience; the individual must consent to it. This is the raison d'être of the contract, which is designed to protect one's life, liberty, and property. Likewise, the church cannot fulfill its mission of saving souls without the express consent of those souls. The consensual, ascending form of the contract takes the place of the descending principle of authority and legitimation (divine right, apostolic Episcopal succession).[34]

For Locke, "everyone is orthodox to oneself." This remarkable statement would dissolve the Christian community into a number of national churches, themselves divided into a myriad of sects, which are in turn divisible into many individuals. The doctrine of consent ultimately atomized orthodoxy into millions of moving grains of sand. It was a great secularizing force that would eventually lock up religion in the cavern of the self.[35]

Locke did not advocate disestablishment of the Church of England, an idea that had barely surfaced during the civil wars. Rather, like the Presbyterians and other dissenters, he seemed to have wanted radical toleration within the establishment itself—in other words, basically the regime that was obtained in the Toleration Act of 1689.

But Locke's implicit aim was to secularize the English state. No longer would it have the role of deciding "indifferent matters" concerning ceremonies and even doctrine. Rather it would be restricted to the civil and temporal well-being of its subjects:

> The Commonwealth seems to me to be a Society of Men constituted only for the procuring, preserving and advancing of their *own Civil Interests* . . . [i.e.,] Life, Liberty, Health and Indolency of Body; and the Possession of outward things, such as Money, Lands, Houses, Furniture and the like . . . and the just Possession of these things belonging to this Life . . . the whole Jurisdiction of the Magistrate reaches only to these Civil Concernments . . . and that it neither can nor ought in any manner to be extended to the Salvation of Souls … [36]

The motive for divesting the state of its religious role is, of course, to avoid persecution and protect individual choice. For Locke, a purely secular state would have no authority over the church, which would become a wholly voluntary and spiritual body. Catholics were to be excluded from toleration, and Jews were to be left alone. Just as the Jews themselves tolerated the idolatrous Moabites and refrained from making converts by choice, so also they would be tolerated in England. What is more,

> If we allow the *Jews* to have private Houses and Dwellings amongst us, Why should we not allow them to have Synagogues . . . Neither *Pagan,* nor *Mahumetan,* nor *Jew,* ought to be excluded from the Civil Rights of the Commonwealth, because of his religion . . . But if these things may be granted to *Jews* and *Pagans,* surely the condition of any Christians ought not to be worse than theirs in a Christian Commonwealth.[37]

Catholics were evidently not Christians. They and atheists were to be excluded—the first because of their reputed allegiance to a foreign prince, and the second because "promises, covenants and oaths . . . have no hold" upon them. On the latter, Locke differed from Pierre Bayle, who, in 1697, wrote in his *Historical and Critical Dictionary* that a society of atheists would not necessarily be immoral, because humans do not act according to their principles. Locke, on the other hand, in his *Essay Concerning Human Understanding,* stated unambiguously that the existence of God could be demonstrated by reason, and that therefore its acceptance must be expected of all men. He concluded that since atheists "undermine and destroy all Religion," they could have no claim to toleration."[38]

Locke's statements about what constitutes orthodoxy are radical, because they deny its very existence. Secularization here means the retreat of religion from the public to the private sphere. The state would become more secular, while the individual, who might privately become more religious, would lose his public voice. This discrepancy between private and public professions of faith constitutes an important dimension of Locke's secularism.

### iv

In April 1690, a critic of Locke's religious views, Jonah Proast, a one-time chaplain of All Souls College who had been expelled from Oxford, published an anonymous critique, *The Argument of the "Letter Concerning Toleration" Briefly Consider'd and Answer'd.* Locke answered, anonymously, the following month with *A Second Letter Concerning Toleration* in defense of "the author" of the *Letter,* and this exchange would continue to a total of four letters from Locke responding to three from Proast. Locke's letters totaled more than 500 pages. (His last years were also spent corresponding with his Continental and English friends, often about religious matters, and on his extensive *Paraphrase and Notes on the Epistles of St. Paul,* published posthumously. This author of major works on epistemology, education, and politics took religion extremely seriously and continued to do so as an early modern secularist who accepted pluralism.[39]

Proast's main disagreement with Locke was that Locke valued toleration more than doctrinal conformity, while he, Proast, believed that some measure of coercion —"indirectly and at a distance"—was in order. Locke was unconvinced that the Church of England, or any other church, embodied the truth fully enough to warrant forcing people to enter into or remain in it. In his *Third Letter,* he focused on the credible or incredible claims to truth of the church in question, not on the rights of conscience of the coerced individual. His strongest argument was that religious coercion produced the opposite of the desired results—a pragmatic rather than a principled decision.[40]

The problem, as Locke saw it, went back to the problem of infallibility. How can one be compelled to accept the teachings of any of the churches if they are not inerrant, or infallible? On the other hand, the

> church of Rome which pretends infallibility declares hers to be the only true way; certainly no one of our church, nor any other, which claims not infallibility, can require any one to take the testimony of any church, as a sufficient proof of the truth of her own doctrine. So that true and false, as it commonly happens, when we suppose them for ourselves, or our party, in effect signify just nothing, or nothing to the purpose; unless we can think that true or false in England, which will not be so at Rome, or Geneva and vice versa.[41]

In order to make a general case for toleration, Locke is led to deny the existence of complete truth in any church in any place.[42]

Locke's dogged and repetitious development of this argument is as much individualist as it is skeptical, for he shudders at the thought of forcing people to come to any church that does not hold the truth. But then, by what authority can Locke claim a hold on truth?

The components of his secularism are sociological (men do not agree on one religion), philosophical (no one denomination can be proven true by reason), and individualist ("everyone is orthodox to himself"). At best, a denomination could come to a makeshift agreement on some fundamental tenets. But, even here, the difficulty for Locke was his questioning such basic dogmas as original sin, the divinity of Christ, and the Trinity.[43]

Locke presupposes that man, left to himself, will disinterestedly pursue truth rather than falsehood. This supposition would seem to spring from a greater belief in the natural goodness of man than Locke actually had, as is evident from the aspersions he casts on his opponent's reasoning, which he calls "false and detestable" or a "fallacy . . . too gross to pass upon this age."[44]

And if a false religion does *not* have the right to use force, does a true one *have* it? This question seems to logically flow from his position. But he denies it, asserting that the original true religion, Christianity, did not need to use force to spread in its first three hundred years. "A religion that is of God wants not the assistance of human authority to make it prevail."[45] These are definitely his most winsome words on religion.

Locke, the philosopher of individualism, naturally invokes the rights of conscience, which he formerly felt the sects abused. Individuals should follow "their own consciences . . . in matters of their own salvation, without any desire to impose on others." If men are left to their consciences, "true religion will be spread far wider . . . than ever hitherto it has been by the impositions of creeds and ceremonies . . . The care of every man's soul should be left to him alone." Again, this view expresses an especially un-Calvinistic confidence in the goodness of man, and his readiness to recognize and accept true religion when he sees it. Yet the very multiplication of sects belies this optimism.[46]

Locke presides magisterially over his disintegration of Christian orthodoxy. Indeed, he often sounds like a relativist: "You, who are in the right way in England, will be in the wrong way in France. Everyone here must be judge for himself." "Mass in France is as much supposed the truth, as the liturgy here." The (atomistic) truth about religion is that there are scarce "3 considering men . . . who are in their opinions throughout of the same mind." A church is like an individual, "orthodox to itself in all its tenets." The irony is that this too is a dogma—a modern one.[47]

Coercive power, granted to national or "confessional" churches, he goes on, achieves only an outward conformity and hypocrisy, not the true religion; "personal gain" and preferments; and "some poor secular advantage," but not piety. The one apodictic truth he admits is that any church that claims to be the true one is most assuredly wrong. To undo dogmatism, he thought, it was necessary to advance a counterdogmatism. The public will always sympathize with the persecuted, and the persecuted will always hate the persecutor and his religion. Furthermore, there is no end to the escalating force necessary to achieve submission; the persecutors themselves will fall into the "pit of perdition."[48]

*Reductio ad absurdum* is Locke's last weapon: he thinks it would be more logical to punish stammerers to stop them from swearing than to punish recalcitrant unbelievers for their unbelief. Locke, who seriously curtails free will in his *Essay Concerning Human Understanding,* here emphasizes that man cannot help being what he is. He is infallible because he cannot be otherwise.[49]

Locke concludes with two players, the secular state and the infallible individual. In his four *Letters Concerning Toleration,* he crafts an utterly secular role for the state and grants no "coactive power" or "plenitude of power" to any church.[50]

The corpuscular individual cannot be reconciled to the national religion, to which all subjects are bound to belong on pain of amends, penalties, deprivation of civil rights, and exclusion from office—resulting from the English Test Acts, France's Revocation of the Edict of Nantes, Geneva's Calvinist consistory, Spain's Inquisition, and Germany's confessional principalities.[51]

Not only was Locke defending his *Epistola de Tolerantia* at length in these years, he was also attempting to reconcile revelation with reason. The central message of the Gospels, for Locke, was that Jesus Christ was the Messiah promised and sent by God. Christ's life was accompanied by miracles, which, Locke maintained, were unique to Christianity and therefore gave it a special position in the panorama of the world's religions. He portrayed the gentleness, benevolence, and humanity of Jesus in a very sensitive way, asserting that Christ died for mankind.[52]

Locke was drawn into a related pamphlet controversy, this time with John Edwards, for omitting mention of the Trinity, original sin, and an explicit statement about the atonement of man's sins by Christ's death in his *Reasonableness of Christianity*. In his *Vindication of the Reasonableness of Christianity* (1695), Locke answered that he should not be attacked for things he omitted. But Edwards only continued with more broadsides alleging Locke's Socinianism (anti-Trinitarianism). Locke *did* minimize original sin by denying that it entailed hereditary guilt, and he was accurately criticized for not explaining what he meant by the term "Messiah" or the term "Son of God." As early as the 1660s, he argued that the Trinity was a doctrine that could not be proven. But nowhere does he expressly deny the Trinity.[53]

Locke's secularism did not consist of irreligiosity—not even the irreligiosity that he was accused of by his Anglican opponents. In August 1703 he wrote to his cousin Peter King:

> You ask me, What is the shortest and surest way for a young Gentleman to attain a true Knowledge of the Christian Religion, in the full and just Extent of it? . . . to this I have a short and plain Answer: Let him study the Holy Scripture, especially the New Testament. Therein are contained the Words of Eternal Life. It has God for its Author; Salvation for its End; and Truth, without any mixture of Error, for its Matter.[54]

By any standard, it is an explicit, private profession of faith.

In the last years of his correspondence, Locke became close to the Socinians and their friends, such as the archbishop of Canterbury, John Tillotson (1630–1694), the anti-Calvinist Remonstrants in Holland, and deists such as Anthony Collins (1676–1729). Locke was very much committed to a lay interpretation of Scripture, disparaged the clergy more than once as the "cassocked tribe," and was fascinated by the number of heretics that England had burned at the stake since the Reformation.[55]

Focus on the Bible did not necessarily preclude a secular mentality, at least not in the early modern days. Locke was totally independent of and even at odds with the Protestant clergy over his scandalous suggestions such as whether matter can think. Some major sources of the Enlightenment were the independent exegetes of the Bible, such as Richard Simon (1638–1712) and Pierre Bayle (1647–1706), and deists, such as John Toland (1670–1722) and Matthew Tindal (1655–1733). If the Bible was not what the churches said it was, then the whole scaffolding of Christianity could collapse. What seemed rare in the beginning of the Enlightenment was a lay, but orthodox, biblical exegete. It was for this reason that the Catholic authorities, and sometimes the Protestant as well, continued to exercise such tight control over the translating, reading, and interpreting of the Bible. The translation of Locke's friend Jean Le Clerc was, for example, banned in Calvinist Holland. The activity of Locke and his friends was secular in so far as it represented a privatization rather than an establishment of religion.[56]

Locke's *libido sciendi* was by no means restricted to biblical issues in his last years. The world of the early Enlightenment emerged in the last volumes of his *Correspondence.* He was interested in the putative Jewish origins of the doctrine of the Trinity, in the reformation of the calendar, in the kabbalah, in comparative theology, in the Jesuits, and in Confucianism. The whole Orient opened up for him now, as a commissioner of trade, as had the Occident earlier in 1668, when he served as secretary to the Lords Proprietors of Carolina. His books were being translated into European languages—Dutch, French, German, Latin—and would reach more and more of the West as the Enlightenment progressed, as is evident through the large collection of eighteenth-century editions of the *Essay* in Harvard's Houghton Library. As for the future, he felt that the end of the *saeculum,* of time, and the beginning of an afterlife would not arrive until the Second Coming.

v

John Locke, a secularist? How can that epithet be tagged on a man who believed in so much of the supernatural creed of Christianity, read the Bible seriously, and counseled others on how to study it? The answer lies in Locke's compartmentalizing of the great spheres of human activity. After his confused early essays on toleration, church-state affairs, and infallibility, he focused on one question that determined all the others. How could anyone impose his belief on anyone else when he could not be sure that his own creed or church was the true one? All the churches contradicted one another. Rather than conclude that *all* were wrong except one, Locke concluded *none* of them was right. The political counterpart was the contract theory wherein no one could command without the other's consent.

Concomitantly, Locke denied that revelation or any church doctrine could be proven by reason. But he never gave up faith. His *Reasonableness of Christianity* maintained that belief in Christianity was reasonable but not susceptible to rational proof. He believed the existence of God was demonstrable by reason (another reason why atheists were intolerable), but knowledge of God was a matter of faith. This last was a fairly modern position shared by many contemporary continental skeptics. Since he believed that the bulk of Christian belief was unprovable, it could not be mandated by church or state. He therefore resorted to what he judged to be the better of two alternatives: toleration rather than infallibility.[57]

Infallibility leads to the second major aspect of Locke's secularism. By the time he wrote his *Epistola* of 1689, he wanted to define the state agenda on purely nonreligious terms: it should deal only with the civic and material concerns of mankind and should not meddle in religious matters, especially doctrine, not even ceremony. This, if adopted, would mark the end of the confessional state and the established church. Locke, characteristically prudent, did not spell this out, or act on it himself, but evidently remained a member of the Church of England.

The third aspect of Locke's secularism is his individualism. The individual should be left free to pursue truth. This did not mean a philosophical or religious solipsism—Locke did believe in the existence of a truth for all men. The task was to find it; in this Locke was quite different from the postmodernists, who seem to believe that truth does not exist.

Protestant and Catholic governments, of course, had been attempting for over a century to control the centrifugal forces of sectarianism endemic to the Reformation. All the problems Locke dealt with stemmed from this reformation: individualism and authority in church and state; reason and faith; individual conscience and infallible authority in the interpretation of Scripture; individualism and absolutism in politics; and certitude and skepticism in philosophy of which neither England nor Protestantism had a monopoly.

In respect to Christianity's "reasonableness," he searched for an ecumenical minimum on which most could agree. Ironically, the Oxford scholar, who began his literary career by decrying sectarianism, ended up allowing for it broadly. Certainly, Protestant pluralism often led to a more intense religious experience, like the Methodist Great Awakening, than did orthodoxy. But this same pluralism precluded a unified, public religion. The vacuum was filled by the state: a secularism of the public sphere. For Locke, the philosopher, the individual was the ultimate reality of human existence, and everything must be built around him. This essentially dissolved not only the divine right of monarchy, but also that of community or church, the two pillars of Christendom. In its place came the consenting individual, eager to find happiness in consensual politics, commerce, and consumption. The secular *république des lettres* or "party of humanity" was founded—an unofficial international fellowship of men of letters who sought liberal learning independently of church, state, and social orders. This increasingly secular *république* would become the vanguard of the eighteenth-century Enlightenment. Meanwhile, traditional Christian civilization continued to unravel.[58]

# Chapter Eight
# Rousseau: The Secular Hermit

Locke, as we know, sought a secular state that would tolerate all religions—save Catholicism—so that humankind could get on with life's more important issues, such as the acquisition of property and the pursuit of knowledge. When Jean-Jacques Rousseau (1712–1778) reached Paris in the early 1740s, he had imbibed Locke's sensationalist psychology and his social contract theory of government. But that hardly accounts for the French philosophe's thought. Adorer of nature but spoiled by salons, fanatic of natural goodness who was convinced of a malevolent plot against him, friend of the Enlightenment who opined that the cultivation of the sciences and the arts corrupted humankind, champion of the Gospels whose own works were burned as heretical, advocate of dualism who is now described as a "materialist," prophet of the people who hobnobbed with dukes and princes, and promiscuous defender of marital fidelity, he was a man whose confessions spoke honestly, yet he was ensnared by complexity—a secular hermit whose retreat spoke less of a love of nature than of an aversion for society.[1]

Hundreds if not thousands of authors have wrestled with these paradoxes of Rousseau's personality and thought. Most of them consider him to have been religious. But how could his religiosity, shaped by Genevan Protestantism, Savoyard Catholicism, and Parisian deism, coexist with Enlightenment secularism? Did his religiosity suffuse all aspects of his life and work, or did he insulate parts of his life from its influence? Alternatively, was Rousseau's religiosity itself secular—aimed at serving purely immanent, worldly objectives?[2]

Rousseau's secularism is less obvious than that of Voltaire, who loathed what he called *l'infame* ("the wretch," or Catholicism), for Rousseau was in and out of the Enlightenment, first accepting its hatred of religious fanaticism and then hating its own fanaticism. He rejected the Parisian philosophers after 1756, was denounced by the archbishop of Paris in 1762, and simultaneously alienated himself from his native Genevan pastors. Along the way he was obsessed with an alleged plot against him in France and so escaped to England, where his paranoia was only aggravated. In the last years of his life (1770–1778) he returned to France, where he finished his *Confessions,* winning an even greater following after death.

### i

Rousseau had a troubled religious upbringing. Days after his birth, his mother died—something for which he sometimes felt responsible. He was raised in Geneva by his father and by relatives until he was sixteen. The family owned a fine house, inherited by his mother whose family descended from French Huguenot (Calvinist) refugees of the 1550s. Both parents' families were "citizens of Geneva," a title enjoyed by only one quarter of the 20,000 inhabitants. Rousseau's father was a watchmaker, but not a terribly successful one, even though Genevan watchmakers were beginning to acquire their modern reputation. Jean-Jacques was apprenticed to an unsavory engraver. One Sunday night in 1728, he came home and found the city gates locked, which induced him to leave and begin a long vagabond existence. He migrated to Annecy, Savoy—the seat of St. Francis de Sales, marked by the conversion of many Protestants. One of his first visits was to the local priest who referred him to a young widow and notable Catholic convert, Mme Francoise-Louise de Warens. In April, a month after his departure from Geneva, he converted to Catholicism in Turin under the influence of Mme Warens and the bishop of Annecy. He returned in 1729 to Mme Warens, who became his protector, and he in turn, incongruously, her seventeen-year-old lover. Remembering this idyllic life in his *Confessions,* he wrote,

> All the objects which I saw before me seemed to guarantee my approaching happiness. In the houses I imagined rustic feasts; in the meadows, playful games; along the water . . . bathing, promenades, fishing; on the trees, delicious fruits, under their shade, voluptuous tête-à-tête; on the mountains, tubs of milk and cream, a charming leisureliness, peace, simplicity, the pleasure of going without knowing where . . .[3]

Rousseau's depiction of this voluptuous, bucolic existence rivals Diderot's Tahiti.

Rousseau wandered, tutoring in the houses of high-ranking aristocrats, becoming, in Lyons, the tutor of the Mably family (two of whose offspring were the future abbés Mably and Condillac, leaders of the Enlightenment).

His teaching experience provided him with grist for his later educational treatise *Emile* (1762). He had himself learned to read at age three; his father had read classics to him from the family library—Ovid, Plutarch, Bossuet, and La Buyère, among others. Moreover, a Calvinist pastor had tutored him at home. But Rousseau was essentially an autodidact. In 1742 he left Mme Warens for good, for she had taken another lover. He reached Paris the same year, met Diderot, and began his most constant profession—copying music—which may have contributed to the cadence and rhythm of his prose. In 1743/44 Rousseau was the secretary to the French ambassador in Venice, promenading along the Grand Canal in the ambassador's gondola, attended by an entourage in livery and presumptuously taking offense when he was not allowed to sit with the ambassador at a state dinner. It was here that Rousseau seems to have become obsessed with inequality and to have begun his *Institutions politiques,* which became the basis of his published political writings. When he returned to Paris, he did so with a greater understanding of the politics of rank and inequality. He then began to live with his lifelong companion and eventual wife, Thérèse Levasseur. In 1754 he cast off Catholicism and rejoined the Calvinist congregation.[4]

## ii

Rousseau's private revelation occurred in July 1749 as he was reading an announcement of the Academy of Dijon, on his way to visit the imprisoned Diderot in Vincennes. Suddenly he saw "the contradictions of the social system . . . I regarded my century and my contemporaries with scorn when . . . I felt my mind dazzled by a thousand lights . . . [I understood] that man is naturally good, and a host of other great truths enlightened me . . . in a quarter of an hour."[5]

Rousseau's *Discourse on the Sciences and the Arts* (his successful prize competition entry in 1750 at the Academy of Dijon) announced that civilization had been detrimental to humankind because all societies that cultivated letters and the fine arts, rather than the martial ones, became enervated and effeminate. The more society progressed in civility and luxury, he argued, the more it regressed morally. Sparta, not Athens, was the country to imitate. Apart from a silly footnote approving the burning of sacred books, Rousseau alluded to religion very little in this essay besides charging that the "reign of the Gospel" caused greater harm with printing than ancient civilization had without it. Indeed, the *First Discourse* flies in the face of the Enlightenment, arguing, facetiously perhaps, that learning has been more detrimental than useful to humankind. From the standpoint of morality, there had been no upward curve in the history of civilization. Not only had it not been marked by the growth of virtue, as the philosophes maintained, but, rather, Rousseau saw it tainted by amour propre—vanity, conceit, and alienation. If Rousseau were serious, he was questioning the entire philosophe experiment. For, if knowledge were harmful, what was the point of the Enlightenment?[6]

Rousseau could have based his argument on the history of Christianity as had Voltaire, who wished to "crush it" as fraud and fanaticism, but he bypassed this opportunity and attacked civilization generically. Rousseau's stratagem was wholly secular rather than anticlerical.

Five years later, with his reputation significantly augmented, Rousseau competed with an equally dangerous thesis—the "Discourse on the Origin of Inequality," or the *Second Discourse*. He rejected the old regime's elaborate social hierarchy and assumed equality rather than inequality as the desideratum. He proceeded to argue that society, not just the sciences and the arts, was the root of human evil. Savages (he was among the very first to reflect on primitive or "quadruped men") were neither good nor bad, neither moral nor immoral. It was only on entering society, through the division of labor and the enclosure of property, that men were corrupted. Again the culprit was amour propre, that damaging self-love filtered through the eyes of society whose esteem man treasured more than his own. Accumulations of property could stem more from vanity than from greed—a means of impressing others more than satisfying an artificial need. Humankind could lose its freedom in the process and become subservient to, if not the slave of, the most propertied.

In the *First Discourse* Rousseau asserts that civilization is false glitter, not virtue. In the *Second Discourse* he claims that social climbing is deception, or false consciousness. (The satirist Jonathan Swift described it as men climbing on the backs of other men in order to show their derrieres to those below.) Property achieved this goal as well as position did. Rousseau also believed that the first man to put a fence around his land established private property, and hence inequality.

Rousseau, Voltaire said, wanted men to crawl on all fours. It is striking how little Rousseau refers to religion. Genesis is not mentioned in this secular "hypothesis," of the *Second Discourse,* containing his evolutionary account of man's descent. Rousseau portrays Christianity as merely something

Europeans imposed on the natives who, like the Hottentots, rid themselves of it eagerly when they return to the Cape of Good Hope.[7]

There is no Creator, golden age, or Garden of Eden in the *Second Discourse*. Both Pufendorf and Locke had made use of man's original relationship to God as the source of his dignity and his rights. Rousseau dismisses such considerations because of the scanty knowledge we have on the point. "Comparative anatomy has as yet made too little progress," he wrote, to invoke any "supernatural gifts he could have received." Man is bereft of the divine and not yet given a rational account of his humanity. The impasse could be termed decidedly secular in its groping for science and religion, without grasping either. But the scientific explanation is a stronger attraction for Rousseau, and it allows him to ruminate on naturalistic hypotheses. It is a proclivity that has led Mark Hulliung to call Rousseau "materialistic."

Rousseau's first writings on social issues are among his most widely read, and most hotly disputed. They bear on the issue of secularism by speaking to a civilization in purely secular terms. Secularism is not the negation of religion, but a more subtle and complex exclusion of and substitution for it. Modern sociology, anthropology, and politics here replace religion's account of man, creation, and immortality. Where the Bible (Genesis) begins with perfection, innocence, and knowledge, followed by personal sin, collective guilt, and punishment—all reversed by Christ's redemption in the New Testament—Rousseau starts with premoral innocence, followed by social or collective sin (amour propre and property), by a morally regressive and degenerate civilization, and finally by a secular redemption or social contract.

Thus Rousseau rejects the Christian historical trajectory of the "happy fall" that necessitates the Redemption. For him the primitive past is a focal point. But he also lampoons modern secular "progress" because it has introduced and exacerbated inequality. The ideal intermediate state between the primitive and the civilized seems unattainable at this point. The solution is left for some later work.

Rousseau's most pertinent statement about religion is found in the *Second Discourse*, with his famous counterdogma of Vincennes: man is born free and is naturally good; the corollary is that evil arises from social causes alone, that is, from living in society. Therefore it is society that needs changing, not man, for man will be regenerated only by a *social* reformation that is not solely due to his own efforts. This solution is certainly suggestive of Helvétius's environmentalism, which he vigorously criticized later. It is also highly colored by his personal experience.

> Never did I feel perfectly happy with my neighbor or myself. The world's tumult dazzled me, its solitude bored me. I needed constantly to change positions and I never really felt well anywhere.[8]

In his *Confessions* Rousseau acknowledges that he "floated ceaselessly from one thing to another." Again in the *Rêveries,* he says,

> Thrown from my infancy into the whirlpool of the world, I learned from early experience that I was not made to live in it, and that I would never reach the state of which my heart felt the need.[9]

He tried to get a grip on the frenzy of entertainment in Paris by going only twice a week to the theater, an institution he would later condemn in his *Letter to D'Alembert on Theaters* (1758). But frustration with the world continued. Even the country could be spoiled by society.

> Although for some years I had been going fairly frequently to the country, I went almost without enjoying it, and these trips, which were always made with pretentious people, and were always spoiled by the trouble only sharpened my desire for rustic pleasure, the image of which I glimpsed . . . only to feel the more deprived. I was so bored by fountains, arbors, flower beds and the most boring showmen of all that; I was worn out by brochures, harpsichords, *Trio,* knotting silk, stupid witticisms, tasteless simpering, little story tellers and large suppers, that when I glanced out of the corner of my eye at a simple thicket of pines, a hedge, a barn, a meadow; when I sniffed the aroma of a good chervil omelet in crossing a hamlet; when I heard from a distance the rustic refrain of a lace worker's song, I sent to the devil both the rouge and the flounces and the ambergris, while regretting the housewife's dinner and the local wine from the vineyard, I would have willingly slapped the face of Monsieur le Chef and Monsieur le Steward, who made me dinner at the time when I supped and supper at bed-time; but above all messieurs the lackeys, who devour my morsels with their eyes and under pain of dying by thirst, used to sell me doctored wine from their masters' ten times more expensively than I would have paid for a better one at the cabaret.[10]

The world of formal soirees and formal salons overlooking formal gardens suited Rousseau as well as swaddling clothes suit a rambunctious baby. He wished to liberate both.

Feeling the forces of nature well up, he slept about with no apology. His acknowledgment of abandoning his five children by Thérèse Levasseur is startling to unsuspecting readers of *Emile,* but he considered it far more moral than birth control and abortion. His rationalization was that by depending on aristocrats who offered to bring up his infants, his children would mix in a society of obsequious lackeys (like Diderot's *Rameau's Nephew*), fawning favors of the rich and the powerful. By repudiating a delegitimized society, rather than taking charge of his illegitimate children, he assumed the high moral ground. He was not marked

> with the depravity of one [who] crushed unscrupulously under his feet the sweetest of duties. No, I declare that is not possible. Never for a single moment of his life was Jean-Jacques able to be a man without sentiment, without heart, a negligent father . . . I will be content in saying that in handing over my children to public education, for want of being able to raise them myself, in destining them to become workers and peasants, rather than adventurers and seekers of fortunes, I believed that I performed an act of a citizen and father and considered myself as a member of Plato's republic.

Rousseau's "confessions" at this point were both unusually frank and heavily glossed by rationalization. Thérèse "obeyed, crying in anguish" at his decision. She had no say in her children's fate; she just wailed. Rousseau, moreover, did not feel that he would have been a good father. He was unworthy of his children, he insists. Besides, if the children were to become aristocrats, he went on, they would reject him. So he abandoned them with *sentiment*—the hallmark of Rousseauesque virtue. He typically blamed the old regime for its lack of virtue. He compared his action favorably with the ancient practice of child exposure, of which there were thousands of contemporary cases yearly. The abandonment of his children was "si bon, si sensé, si légitime." His respect for the natural sexual act flew in the face of the aristocracy's increasing recourse to contraception.[11]

Relations with the world did nothing to quiet Rousseau's restlessness. He had basically three ways to achieve peace. He could retire to nature. He could reason in the manner of the "Savoyard priest," the persona who enunciates Rousseau's natural religion in his *Emile* (1762). Or he could attempt to create anew the whole political, moral, and religious world through the "social contract." He does all three.

> A great revolution . . . just took place in me, another moral world which unveiled before me the senseless judgments of men of which . . . I began to feel the absurdity. [I perceived] the ever growing need of another reward than literary kudos of which the aroma had barely reached me when I was already disgusted by it . . . This [1756] is the epoch at which I can date my entire renunciation of the world and [the arousal of] this lively appetite for the solitude which has never left me since that time.[12]

Rousseau's flight from society to nature was also a flight from organized (i.e., socialized) religion. His "entire renunciation of the world" was not monkishness, but a "secular hermitage." *Emile* was written in the Valley of Montmorency, far from the madding crowds of Paris. There Rousseau proclaimed, "I only began to live on the 9th of April 1756."

Rousseau, in short, came to believe that morality sprang from man's nature as it is, not as it should be:

> The fundamental principle of all morality about which I have reasoned in all my writings . . . is that man is naturally good, loving justice and order; that there is no original perversity in the human heart, and that the first impulses of nature are always upright.[13]

### iii

Self-rule or autonomy struck at the very roots of the old regime. Perfected by Louis XIV, this regime consisted of a personal royal governance by an authoritarian ruler who called himself the "Sun-King," a god on earth. He housed the thousands of privileged noble families who had lost much of their independence at Versailles. He persecuted autonomous religions, and his centralized administration aspired to direct all provincial and cultural life. His police supervised opinion.

Immanuel Kant believed that youth was subject to authority, unable to think, yet, for itself. He defined Enlightenment, by contrast, as an adult's daring to know by himself. Rousseau's Emile has a tutor, but he is taught how to learn by experience, never by rote. Nor is he obsequious, or even deferential, to those above. His focus is not on honor but on sincerity. He is absorbed by "the book of nature." In the country, Emile discovers the "rustic simplicity [which] lets [the] passions develop less rapidly," and he learns through the senses rather than through speculation. Precepts are to be kept to a minimum. Body culture, for Emile, is as important as mental development, as Spartans were fully aware. Toughening the body through stoic practices, such as sleeping with inadequate covers on a hard bed, is a secular asceticism, different in purpose from a religious one. Emile is taught not to suppress the passions, but to channel them. Pity, for example, can deflect from senseless lust and lead to sympathy and good deeds. Emile is steered away from the chief source of corruption, which is society.[14]

As Diderot does in his *Encyclopedia*, Rousseau too extols the manual "arts" such as carpentry, which is held up as honorable, not disgraceful, to practice. Emile will not depend upon a retinue of servants, but will wish constantly to do things himself. Self-reliance and experience are encouraged as replacements for faith and authority.

*Emile* summarizes Rousseau's whole social-political philosophy, which is "to obey only oneself," as he stated in the *Social Contract*.[15] St. Augustine opined that the first Christian virtue is obedience. This is one Christian virtue that Rousseau himself resoundingly lacked and of which he deprived Emile, whose independent character was based on self-made rather than received worth, on personal accomplishments rather than hereditary rank. Not only was the dogma of original sin rejected, but also its evidence. Family quarrels over precedence, privilege, and property were attributed to society. Rousseau pointed to the origins of inequality in the first man who put a fence around his land; perhaps he should blame the first child who snatched a twig from his sister's fist. Generosity, like every other virtue, is not natural but acquired.

*Emile* posed the problem of society in terms identical to the *Social Contract*. Compare "Everything is good when it leaves the hands of the Author of things, everything degenerates in the hands of man" (*Emile*) and "Everywhere man was born free, and he is everywhere in chains" (*Social Contract*).[16] Rousseau had to free society in order to free himself from society's amour propre, the secular counterpart of original sin. His own self-love is centered in the publication of his books, in receiving approval from readers, or, inversely, in the burning of those books by his enemies. True autonomy of the self, he felt, was achieved only through the "I" coming to terms with the "other" on one's own terms, rather than accepting the judgment of the other over oneself. Obedience to one's own law is freedom. In Rousseau's words, the self-made social contract means, "No longer ask if the law can be unjust, because no one is unjust to himself."[17]

Rousseau's analysis lies at the heart of the modern secular outlook. Through the Christian centuries, evil was thought to have been inherited from Adam, and the sole hope for overcoming it was through grace dispensed by the church's sacraments. Great epochs of goodness—the apostolic age for instance—occurred because of the merits of saints drawing on those of Christ. The city of God could push back the frontiers of the earthly city, but never completely. The *saeculum* was neutral at best, unredeemed at worst.

Rousseau putatively delivers mankind from this burden of sin, which he himself relocates not in the individual, but in society. The solution is therefore either in retreating from society to a secular hermitage, which he himself enjoyed in the Valley of Montmorency after 1756, or in bringing society under control of the "general will" through a social contract. Such a polity would enshrine complete liberty and autonomy for all. Inequality would disappear because everyone would be the equal author of his own membership.

The *Social Contract* anticipated the French nineteenth-century revolutionary quest for self-determination: elections, referenda, revolutionary *journées,* insurrections, and external campaigns of liberation, none of which Rousseau foresaw or endorsed in advance. Ecclesiastically, it helped unchurch millions who rediscovered God in forest and stream, rather than in dank stone churches. The world of the formal French—their gardens, salons, and academies—is jettisoned for the world of the grove, the gorge, and the waterfall. Sexually, Rousseauianism coincided with a significant increase in illegitimate births—another form of escape from the strictures of church and community. Rousseau's own hermitage differed from that of a monk by his sexual relationships. If marriage were to be saved, à *la Nouvelle Heloise,* it should not crush the heart.

Rousseau's appeal for liberation spilled over into social-political movements such as abolitionism and the democratic social contract. Man is in charge: no hereditary guilt weighs him down, no apostolic succession binds him, no dynastic monarchy commands him, and no immemorial aristocracy subjects him. Rousseau's message to a crippled mankind is, Get up and walk. It is a rejection of Reformation Christianity, Calvinist, Catholic, and Jansenist, for a humanized (secular) facsimile.

### iv

Like most thinkers of his generation, Rousseau rejected the idea of dogma—that is, a body of revealed doctrine interpreted by a church. Dogma was the antithesis of Enlightenment, or learning acquired by human effort and ingenuity. The philosophes sought to subject all such beliefs to rational examination (*libre examen*), firmly rejecting the notion that the longevity of an idea was the measure of its truth. They, like Voltaire and Kant, spurned the idea of a historic revelation.[18]

Like Locke, Rousseau was troubled by the phenomenon that each religious sect considered itself to be the true one; his own alternation between Protestantism and Catholicism probably bred skepticism. But, in general, Rousseau was not bothered as much as Voltaire by the historical record of religion. Rather, he cared about the internal consistency he thought he found in Christianity. He disliked the catechism, probably because it was taught mostly by rote rather than by reason, and he abhorred the Roman Catholic dictum "Outside the church there is no salvation," opining that anyone who uttered such intolerant words should be driven from the state—hardly a tolerant alternative! He himself entertained no hope of knowing the true religion, for, like Locke, he deemed it impossible for an individual to discover it in a lifetime's investigation. The Catholic Church, created for this purpose, could not, he believed, be trusted. Pessimistic about attaining religious truth, he urged his correspondents, who addressed him on this question, to keep the religion of their fathers.[19]

Rousseau clashed with the church on many issues, but one assumed the greatest importance, and that was original sin. Virtually every Western philosopher had something to say on this subject—whether ecclesiastics like St. Augustine who formulated it, Aquinas who limited its effects, Calvin who exaggerated it, or Machiavelli who proved it. Rousseau's objection was that it was essentially unjust to blame men for sins they had not themselves committed. Man, he felt, was good enough to enter society without baptism, for which the social compact was a secular substitute. Rousseau had read, but had not accepted, Malebranche's argument that the presence of moral and physical disorder in the world, so repugnant to God, was the living proof of original sin.[20]

He objected to other Christian claims as well. He examined several miracles, such as the raising of Lazarus from the dead, and concluded that the Scripture's contradiction of reason was due to humans' alterations of the Gospel rather than God changing nature.

All this dissent notwithstanding, Rousseau defended himself to the archbishop of Paris: "I am a Christian, sincerely Christian according to the doctrine of the Gospel." But he did not sincerely accept "the doctrine of the Gospel," as we have seen, or if he did, it was a rationalist's gospel, without the supernatural. What he did was to continue the process of relocating authority begun by the Reformation: Scripture has greater authority than the church; reason has greater authority than the Scripture; and the heart, "the source of true morality," has still greater authority than reason. Thus invoking emotional and individualistic criteria, Rousseau bypassed inscrutable Christian dogmas.[21]

"I often travestied religion in my style, but I was never completely without religion," Rousseau stated in his *Confessions*. If he reacted against dogmatic Christianity, Protestant as well as Catholic, he was not, on that account, going to hurl himself heedlessly into the philosophes' "party of humanity." Sounding almost like an antiphilosophe, he considered the age he lived in to be "the scum of centuries" and distanced himself from "modern materialism," this "sad philosophy." At a dinner at the house of M. Quinault, sometime after the Lisbon earthquake of 1756, when the guests seemed to call Providence into question, he exclaimed courageously in the presence of atheists, "As for me, Monsieur, I believe in God." But far as Rousseau might remove himself from the "fanatical" philosophes, he left the ranks of the "fanatical" Catholics as well. He was quite critical of the good priest who had converted him to Catholicism, because he said he did not estimate him to be a virtuous man. He was equally negative about a homosexual abbé in Lyons. He was not on good terms with the Jesuits, the enemies of the *Encyclopedia,* or with the Jansenists, cavalierly dismissing them for their "atheistical fanaticism" or their "pious fanaticism" respectively. In truth, he disliked all factions, but especially the "materialists" and philosophes who tended to savor every clerical scandal as proof of clerical villainy. He concluded with one salutary thought that the crimes of the clergy did not disprove the religion they represented.[22]

The *Profession of Faith of the Savoyard Vicar,* which forms part of Book IV of *Emile* (1762), features an independently minded country priest who defends natural religion with reason and sentiment, rather than dogma. Peasants, moreover, have saner minds than city folk, observes the vicar, through whom Rousseau probably speaks. Revelation does not reach all mankind—an objection raised gently by Dante and severely by Kant. Even if the Gospel penetrated to the farthest corners of the earth and yet left one man uninformed, it would be inequitable and consequently invalid. "In the end," protested Rousseau, "it is a book; a book ignored by three-quarters of the human race. Am I to believe that a Scythian and an African should be less dear to a common Father than you and I?" Rousseau, and later Kant, wanted a universal and simultaneous revelation that would leave nothing for man to propagate. Rousseau did retain the notion of an intelligent cause of the universe as a datum of reason. Yet he rejected the Augustinian, Jansenist notion of grace and salvation offered only to some. Nor did he accept the papal and Jesuit insistence that Christ died for all mankind. The Gospel, of course, created the enormous responsibility of spreading Christ's message by the words "Go to the whole world and preach this Gospel to all mankind" (Mk 16:15).[23]

Instead, the universality of the Gospel is exchanged by Rousseau for the universality of nature and reason. In criticizing the supposed inadequacies of the Judeo-Christian revelation, Rousseau continually falls back on the Gospel's *moral* as opposed to its *dogmatic* teachings: "The Gospel is alone in its morality, always certain, always true, always unique and always consistent with itself in itself." And again: "All my writings radiate the same love for the Gospel, the same veneration for Jesus Christ." The Gospel morality is to be accepted because it is true, and not considered true because it is the Gospel. Exchange of a dogmatic religion for a moral one was in perfect step with the Enlightenment.[24]

Rational as his negations and heartfelt as his affirmations might appear, they were not enough to satisfy rationalists or Christians. As a result, he got on poorly with the Paris salon philosophes and the Protestant clergy of Neuchâtel and Geneva. Within days, the notoriety of his writings caused his banishment in 1762 from France, where *Emile* had been condemned in June by the Parlement of Paris. Ten days after that, the Little Council of Geneva burned his *Emile* and *Social Contract.*[25]

At Môtiers in Neuchâtel, where he took refuge in July, he persuaded the Protestant minister to permit him to take communion. But his religiosity did not keep him churched for very long. To the archbishop of Paris, Christophe de Beaumont, he wrote a "letter" in November 1762, after Beaumont had issued a pastoral letter condemning *Emile,* particularly its *Profession of Faith of the Savoyard Vicar.* Like Locke, Rousseau spent part of his last years defending himself against the clergy's charges of heresy. Unlike Locke, he descended to ad hominem insult. Beaumont had accused Rousseau of having "set himself up as the Preceptor of the human race in order to denounce it . . . the oracle of the century in order to completely deceive it." Rousseau countered, "Why, Monseigneur, should I have anything to say to you?" Repeatedly and condescendingly he alluded to Beaumont's "error" and accused him of "bad faith."[26]

Rousseau's *Letters Written from the Mountain* (1764) was meant to vindicate himself again in the eyes of the Genevan theocracy, but he continued to remain persona non grata to them as well. Not only did that clergy repudiate him, but the villagers of Môtiers stoned his house in 1765 for similar cause. He thus left Neuchâtel and was soon en route to England. There, David Hume tried to host him, but Rousseau's suspicions drove him back to France to seek solitude, after having failed to make up with Catholic, Protestant, and philosophe acquaintances.[27]

How could one who wrote so much with such fervor, about religion be called secular? In the two *Discourses,* the historical, revealed content of religion is largely knocked out, and progress and regression are measured in purely moral terms, such as amour propre, which takes the place of original sin. In Rousseau's major works of the 1760s, especially the *Confession of Faith of the Vicar Savoyard,* religion is esteemed, but it, too, departs from Catholic, Protestant, and, yes, philosophe orthodoxy. True, the "confession" is put in the mouth of a priest, but one who is peasant born is excused for becoming a priest, because he bent to his parents' wishes. The philosophes generally hated the hierarchy and religious orders, but could esteem the parish priest close to the people. The "bon curé" was a "trope" of Enlightenment polemics. Yet the Savoyard's defense of his faith is not entirely credible, for Rousseau puts in his mouth a fairly sophisticated philosophical apology, not for the Catholic Church, but the truth of "natural religion," which the curé has explored, finding the skeptics' doubts about it quite wanting.

> Imagine all your ancient and modern philosophers having exhausted their bizarre systems of force, of chance; of fatality, of necessity, of atoms, of the animated world, of living matter, of every kind of materialism.[28]

Rousseau denies the materialist contention that since movement is inherent in matter, there is no reason to seek the origin of movement outside matter. He asks himself, On what grounds can one judge such things? And he concludes rhetorically with a Cartesian question to end all doubts: "Do I have a sentiment of my own existence?" Consciousness is the foundation of philosophy. Opining that "judging and sensing are not the same thing"—thus taking distance from Helvétius, the materialist psychologist of the 1750s, and laying the ground for a spiritual soul—Rousseau believes that sentiment is man's spiritual faculty on which reason builds. Thus, the choices are "body senses, passions, passivity vs. soul-liberty-reason activity." Contrary to Condillac, he believes that reflection does not derive from sensation but points rather to an "immaterial substance." The materialism of the philosophes is what induces Rousseau to describe the Enlightenment century as *la lie des siècles,* or the "dregs of the centuries." "The modern philosophes hardly resemble at all the ancient philosophers" who believed in God.[29]

This religion of nature and of reason is not entirely new. Its essential elements can be traced back through the Middle Ages to Plato and Cicero. It was what the scholastics called (natural) theodicy and what Leibniz dubbed "theodicy"—previously a handmaid of revealed, dogmatic theology. Not only was it not subversive, it was the indispensable human foundation for accepting Revelation. However, for Rousseau, it was no longer the handmaid, but the mistress. At the apex, natural religion pretended to self-sufficiency, to the place of theology. This is why the faith of the Savoyard vicar, who is the century's best exponent of this natural religion, is a secular religion, for it is cast in the language of *philosophie,* which at that moment had the last word. There is no mention of Christ or anything Christian in the "Confession," which with minor editing could be placed in the mouth of Aquinas, Aristotle, Plato, Cicero, Locke, Leibniz, Clarke, or Pufendorf. That is, it is a secular proof of God's existence and governance. Rousseau, however, embraced it, like many reformers of the old regime, because it cleared out the ecclesiastical and baroque trappings of eighteenth-century Catholicism. The Jansenists were acting similarly within the church by substituting austere ritual and liturgy for the Jesuit baroque. Latching onto "natural religion," Rousseau created a benevolent and rational God, whose discovery through sentiment and conscience undercuts the atheistical tendencies of the materialistic wing of the Enlightenment—the famous "Holbachic clique." Thus, he made room for the élan of his soul while cutting free from the clergy and the philosophes. He obeyed only himself in keeping with modern natural law individualism.

Rousseau owes a lot to English deists like Matthew Tindal (1655–1733), who wanted natural reason to supersede revelation whenever there was a contradiction between the two. He was among the first to turn natural reason against Christianity, of which it had previously been the support. More orthodox theologians whom Rousseau also read were Samuel Clarke and Gottfried Leibniz, who corresponded famously about the compatibility of the "new philosophy," Newtonian physics, and natural religion on the one hand, and Christianity and Providence on the other. The idea could be also found in Christian von Wolff. For all three, Newtonianism did not nullify the Christian Revelation.[30]

### V

Where do church and state come in? Rousseau wrote almost four books of the *Social Contract* without mentioning religion. He did devote his last chapter, however, to "civic religion." In antiquity, Rousseau explains, each tribe or city had a religion and did not distinguish between "its Gods and its laws." There were no religious wars because "the Gods of the Pagans were in no sense jealous Gods." The trouble began "when the Jews . . . stubbornly sought to recognize no other God but their own." Rousseau tolerated Jewish exceptionalism no more than Christian exceptionalism.

It was in these circumstances that Jesus came to establish a spiritual kingdom on earth; this kingdom, by separating the theological system from the political, meant that the state ceased to be a unity, and it caused those intestine divisions that have never ceased to disturb Christian peoples. Now, as this new idea of a kingdom of another world could never have entered the minds of pagans, they always regarded the Christians as true rebels who, under the cloak of hypocritical submission, only awaited the moment to make themselves independent and supreme and, cunningly, to usurp the authority that they made a show of respecting while they were weak. Such was the cause of the persecutions.

> What the pagans feared did happen; then everything changed its countenance; the humble Christians changed their tune and soon the so-called kingdom of the other world was seen to become the most visible and violent despotism of this world.[31]

This is not Locke, who believed that the early Christians' lives and miracles were exemplary testimonies to the truth of Christ. Nor yet Gibbon, who thought that the early Christians' meekness and aversion to the martial arts were partly responsible for the collapse of the Roman Empire. Rousseau's argument is based on the phenomenon of division of loyalties between Caesar and Christ, which Christianity effected in the undivided ancient city—a double allegiance, one to the church and another to the state: "Two legislative orders, two rulers, two homelands, puts them under two contradictory obligations, and prevents their being at the same time both churchmen and citizens . . . Such is Catholic Christianity." That church, in his mind, will always be the enemy of the state.[32]

What about Protestant Christianity? Rousseau invokes it approvingly, if briefly, by referring to Hobbes, who would put the church under the aegis of the state, "reuniting the two heads of the eagle and fully restoring that political unity without which neither the state nor the government will ever be well constituted." In the *Letters from the Mountain* Rousseau clarifies this: "In the Principles of Protestants, there is no other church than the state, and no other ecclesiastical legislator than the Sovereign." This would have pleased Marsilius of Padua, but it rejects the medieval dualism of Aquinas, Dante, Becket, and St. Louis.[33]

Rousseau, however, does not favor returning to the original religion of the ancient cities of Greece and Rome: "To the one nation which practices this religion, everything outside it is infidel, alien, barbarous; and it extends the rights and duties of man only so far as it extended its altars." Thus he withdraws his introductory enthusiasm for the ancient cities' civil religion and thereby becomes a critic of "ancient liberty."[34]

In the end, after all his mental twists and turns, Rousseau seems to believe that religion is indispensable to any state, but that all historical and existing religions, with the possible exception of Protestantism, have been harmful to it. The ideal religion will have dogmas, those of deism just enumerated, which a citizen must believe or be banned. Is this not a secularization of the Catholic *extra ecclesiam nulla salus est* ("outside the church there is no salvation")?[35]

What are the secular dimensions of this chapter on "civil religions" in the *Social Contract?* It disestablishes all existing Christian churches by granting all an equal footing. The new civic religion would be immanentist, as opposed to transcendental: it would exist for the sake of the earthly needs of the community created by the social contract. The otherworldly dimensions of the civic religion exist only to supplement human justice. Such social justification of religion plays a prominent part in different ways in the thought of Voltaire, Necker, Robespierre, Napoleon, and Chateaubriand. In short, the raison d'être of religion is conceived as secular.

Rousseauism, along with Jansenism and Gallicanism, inspired the Revolution's civil control over the church. The prohibition of clerical assemblies, especially any linked to Rome, and the termination of clerical control over marriage are two major secular objectives that were realized a generation after the publication of the *Social Contract*. Legislation creating the "civil state" (*état civil*) of 1792 secularized the record keeping of the vital acts of births, marriages, and deaths and struck at the dualism of church and state that Rousseau abhorred. While parishes were allowed to keep parallel records, their exclusive control over these acts of passage was broken and secularized. The system was later exported by Napoleon throughout his empire.[36]

Rousseau's predilection is for what he calls "the religion of man . . . the pure and simple religion of the Gospel, the true theism, which might be called the divine natural law." This is "the religion of the private person, not the Christianity of today, but that of the Gospel, which is altogether different." This religion is concerned solely with "the things of heaven"; its "homeland is not of this world." It is "holy, sublime and true," one in which "men, as children of the same God, look on all others as brothers." It is indissoluble even in death. Rousseau allows this religion to exist *privately,* but not as a civic religion, because it is too otherworldly. It is, perhaps, his Savoyard confession, his own hermit's religion. He weaves a cocoon around this religion of the Gospel, and of the solitary man, outside which is the civic religion of the state. The latter is secularized, but a private sphere inside enjoys a purely spiritual existence. Rousseau and the French Revolution end up with just the state and the individual, without intermediaries, but the latter is free to retreat into a hermetic security.[37]

The French revolutionaries—Bernardin de Saint-Pierre, Robespierre, and many others—were most receptive to Rousseau's deist ideas. De-Christianization, the new cult of nature, and the civic festivals sought to replace Christianity. As much as any philosophe he had criticized, he transmitted the Enlightenment to the next generation. Indeed, a case could be made that he did so more effectively, for he conveyed those ideas in a syncretic way (*philosophie* mixed with Christianity), which made them more acceptable than they would otherwise have been. Rousseau was, for a still Christian populace, the least difficult pill to swallow.

### vi

Rousseau is the guide of the mountain climber, herb gatherer, and fossil hunter; of lovers who want to learn how to follow the virtuous inclinations of their heart; and of all who seek to practice virtue by the light of reason and the inner prompting of their sentiment. The list of works by and about Rousseau fills several hundred pages in a bibliography of romanticism. He became an object of a popular cult at Ermenonville and of an official cult when his remains were transferred to the Panthéon after the Terror. His political theories of representation, of general will, and of leadership were important to crucial political figures of the Revolution, including the Abbé Sieyès and Robespierre. The failure to wed the "religion of man" successfully to a civic religion led to an even more established church after the Concordat of 1801 between Napoleon and Pope Pius VII. And so the civic religion taken partly from Rousseau died out. Rousseau's successors in the nineteenth century—Henri de Saint-Simon, Auguste Comte, Jean-Baptiste Fourier, Jules Michelet, Victor Hugo—abandoned most of the revolutionary pageantry, but kept a secular religion of man, giving it a somewhat more anthropomorphic, humanistic face. The object of cult became an immanent God-man.[38]

Rousseau's cult, particularly its religious side, borrowed heavily from the Western philosophical (theistic) tradition, but was divorced from the Christian churches, none of which was happy with him and with none of which was he happy. He devised a kind of parallel secular morality for marriage, education, religion, and governing. He was deeply indebted to Christianity, yet his penchant for originality would not allow him to actually "subscribe" to it. He was a philosophe who protested against it and a Christian in his critique of the philosophes. He fused the new philosophy with religion and was found more acceptable to millions for that reason. His thinking would secularize the public sphere by excluding the church in the *Social Contract*. But through his championing of conscience, the individual could retreat into himself, by himself. His inconsistencies, far from leading to his repudiation by the following generation, offered something for everyone—even counterrevolutionaries. There was reason for his apotheosis after the Terror in the secularized Panthéon of the new great men of the nation.[39]

# Chapter Nine
# Immanuel Kant's Ambiguous Secularism

The end of the eighteenth century witnessed a crystallization of the Enlightenment in Germany and France, an attempt to organize and formalize it in systematic treatises rather than in satirical dialogues and tales. The Enlightenment, it was felt, was an ongoing process of indefinite perfectibility and would be made more rational by the nature of reason itself.[1]

Kant, like most graduates of secondary schools in the eighteenth century, had a good classical education. But he keenly disliked the school's church services, despite his Pietist background. In 1740, he entered the University of Königsberg, where he stayed on as a professor for the rest of his life. He taught as many as eight subjects, ranging from logic to math to geography, and eventually received the chair in philosophy in 1770.[2]

Intellectual life in German universities was much more active than in France. None of the leading French philosophes of the Enlightenment held a university position; indeed, most lacked a university education. The French Enlightenment grew outside the universities, often in protest against their sclerosis, particularly that of the scholastic philosophy and theology departments. But things were different in Germany. Virtually all philosophical endeavors were carried out in the leading institutions of higher learning. There, too, remnants of medieval scholasticism were evident in the dominant philosophies of Gottfried Leibniz (1646–1716) and Christian von Wolff (1659–1754), who held sway before Kant.[3]

i

Kant never wanted to leave Königsberg, despite an attractive offer from the University of Halle in 1778 that would have made him a neighbor of Goethe and Schiller. But Kant had a world, even a universe, within him, which he had to explain. Moving would distract him from his grand enterprise: "All change frightens me . . . I must obey this instinct of my nature if I am to spin out to greater length the thin and delicate thread of life which the Fates have spun for me."[4] For a critic of reason, the truth was to be found by turning within, not by seeking experience without.

The *Critique of Pure Reason* appeared in 1781. The title reveals it as an Enlightenment work, because epistemology, or the knowledge of how we know what we know, had become the Enlightenment's principal branch of philosophy since the great Descartes had wondered what, if anything, he knew. Descartes had reestablished Continental philosophy with the cogito of self-consciousness. Locke, on the other hand, argued that most of our ideas came from sensation—that is, from the external world rather than from the mind alone. The Enlightenment, then, inherited two epistemological traditions: the idealist and the empiricist.

In 1783, Kant wrote in his *Prolegomena to Any Future Metaphysics* (that is, any tenable metaphysics), "My remembering David Hume was the very thing which many years ago first interrupted my dogmatic slumber and gave my investigations in the field of speculative philosophy a quite new direction." Why David Hume (1711–1776)? A decade older than Kant, Hume was an empiricist in the Lockean tradition, but far surpassed Locke in his skepticism. For Hume, cause and effect were inferences from experience. But just because experience teaches cause and effect, he continued, we cannot infer that such things exist outside experience. We do not, for example, know whether the world has a cause—a "first cause" in, say, the Aristotelian or Thomistic sense. Since natural laws are invariable, miracles, which interrupt them, are impossible. But one can object that natural laws cannot *always* be invariable, because "all time" or "always" surpasses experience! Sense knowledge or experience is not absolute, but only probable. Knowledge of the beyond is a matter of faith. Fideism was frequent in Protestantism, but in Hume it may not have been sincere. Hume's *Dialogues Concerning Natural Religion* (1779) pretty much demolish the comfortable Enlightenment deism based on the common experience of all religions (existence of God, immortality of the soul, and rewards and punishments after death). But they posit the components of Kant's "antinomies," or contradictions of pure reason, examined below. Hume could have been a fideist as he claimed, but today most see him as an atheist, or "the complete modern pagan."[5]

If the attraction of empiricism was that it rescued man from nonage, from immaturity, it also left humanity with little beyond experience. German philosophers were not interested in surrendering such an important part of reality without a struggle.

Kant's *Critique of Pure Reason* embraced both empiricism and metaphysics. Following Locke, Kant agreed that we do not know substances, but only their apparent properties; and all we truly know are *our* ideas. His most famous proposition was that we do not know "the thing in itself" (*ding an sich*). This is not an entirely original position, for Descartes had already questioned the veracity of our perceptions and the existence of objects exterior to us. He, Malebranche, and Berkeley had only resolved the problem by resorting to God, who guaranteed such knowledge. Kant, on the other hand, wanted to solve the problem by examining our faculties alone, without recourse to a deus ex machina. He called his solution "critical philosophy" because it consisted of reason criticizing itself. This was much in keeping with Enlightenment thinking: metaphysics, religion, social and political customs, and the whole ancien régime were objects of criticism. Rousseau's *Discourses* and Voltaire's *Candide* are two examples. Kant's *Critique*s would go to the very core of Enlightenment criticism by criticizing the very faculty that criticizes! This criticism to the second power is also called "transcendental"—a word that denotes German philosophy in general from Kant to Hegel.[6]

In Kantian vocabulary, to "criticize" does not necessarily mean to disparage, but rather to evaluate, to bring under the bar of reason, to take nothing for granted or on faith. Tradition must come under severe scrutiny.

In his first *Critique,* Kant calls into question the very foundation of knowledge and concludes that we cannot know the nature of things that are outside us. All we know are the phenomena we experience through our "sensuous intuition" (a phrase Kant uses roughly like the French use "sensation"). We cannot go beyond phenomena in knowledge of externals, *except* (and this is an important concession) by supposing that externals do exist. Our knowledge is provided by our senses, but it is also limited by them. Man is not as omniscient as he thinks. Kant, without saying so, seems to question the overconfidence the French had in reason.[7]

There is more. Sensations and intuitions are not really knowledge. A blind man given sight will not recognize a horse when he first sees it. He will only see colors and shapes. Something in the mind and experience is needed to associate these colors and shapes with the concepts of mammal, quadruped, beast of burden, and so on. Since the blind man knows nothing of these concepts, he has to link his new visual experiences with his previous audible or tactile experiences. Eventually, what Kant calls the "categories of understanding" intervene to form a synthetic concept of these new impressions.[8]

Thus, Kant believes that we do not know "things in themselves," but only as they exist in phenomena, which we perceive through intuitions and the "concepts" that shape them. He tried to resolve the difference that had divided philosophers since Plato, who emphasized mind or form over matter and sensible appearances, and Epicurus, who emphasized matter or sensation over mind. Aristotle, the most important source for the West, combined both *noeta* (intelligible form) and *aesthesis* (experience from without). The seventeenth-century Cartesians went the way of Plato; the Enlightenment led by Locke and Condillac pursued the path of Epicurean empiricism. Space and time, for Kant, were "subjective conditions [or categories] . . . of the human mind, for the coordinating of all sensible things in a fixed law." Sensation cannot exist without them. The understanding supplies the concept, without which sensibility would be structureless. The two together produce sensuous intuition, or sensation.[9]

Does this focus on the mind make Kant a subjectivist, an objectivist, or a relativist? Kant's epistemology is both subjectivist and objectivist, with a slight accent on the former. His description of knowledge focuses on the knower rather than on the known, the internal rather than the external, and the subjective rather than the objective. But it is objectivist in that he describes a process that originates externally and takes place in humanity in the same way and so has universal validity. These compromises represent Kant's manner of gathering the various strands of Western philosophy into one orchestral statement, which goes a long way toward explaining his immense contemporary and posthumous influence. Nowhere can this be seen better than in his development of what he calls "the antinomies of pure reason."[10]

ii

An antinomy is a contradictory thesis. The term was first used by the Roman (Spanish) grammarian Quintillian. Kant devotes almost half of his *Critique of Pure Reason* to several such antinomies, which cannot be proved or disproved by pure reason. The method is dialectical and harks back to scholastic "objections" to arguments, except that they are not resolved. Kant highlights several: "Casualty in accordance with laws of nature is not the only casualty . . . there is another casualty, that of freedom . . . There belongs to the world, either as its part or as its cause, a being that is absolutely necessary."[11]

In the remainder of the *Critique of Pure Reason,* Kant drives home the nonprovability of these ideas, particularly the existence of God. Kant is digging a void between pure reason and theology. "Dogmatism" maintained the provability of these metaphysical ideas, but Kant denies their certitude, which was accepted by the "dogmatists," by Leibniz and Wolff, and by the whole scholastic-Aristotelian tradition, even the tradition of Descartes and Malebranche. Many Enlightenment thinkers in France were casting doubt on these same concepts. Kant recasts them in a very magisterial and solemn way, elevating them to the level of the most "critical, transcendental" philosophy. But he cannot affirm one side of the antinomies, as that would make him a scholastic or a dogmatist. And he cannot ignore them, either, because he recognizes their supreme importance.[12]

But if reason is so hemmed in and unable to know the "thing in itself," either *in* nature or *above* nature, how can it pronounce so confidently on its own nature and powers, itself being "a thing in itself?" Hegel arrived at the same objection. Presumably, self-reflection in the Cartesian system establishes one indubitable fact—personal existence. But Kant's critique ends up in insoluble antinomies or contradictions that result not in knowledge, but in doubt, the endless turning of the mind upon itself. Doubt is the philosophical precondition of secularism. For how can religion occupy a place in the public sphere if it is riddled with uncertainty?[13]

The concern here is not with Kant's consistency, but with his place in the history of secularism. Kant did not think that philosophy as the handmaid of theology was compatible with the rigors of transcendental criticism. Theology and religion did not fall within the pale of reason, and thus they should be separated from philosophy. But Kant did not wish to undo traditional religion, as did many of the French philosophers. In France, Voltaire declared a war on the "infamous thing" in the early sixties. Three decades before the Revolution, the anti-Christian *Encyclopedia* of Diderot and d'Alembert was selling vigorously and the underground commerce of pornography was also increasing. On the other hand, German Pietism and the Storm and Stress movement, which emphasized authentic emotionality, exerted a stronger grip on the German intelligentsia. Earlier Counter-Reformation Catholicism and Jesuit anti-Jansenism in the first half of the eighteenth century in France often produced unintended and unwelcome results—for example, an alienation from the church and religion in general. Moreover, instead of a political revolution, Germany, it is said, had a revolution of the mind—the dialectical idealism of Kant, Fichte, Schelling, and Hegel. Kant said that his criticism was meant "to make room for faith."[14]

Between the first (1781) and second (1787) editions of the *Critique of Pure Reason* and three years before the *Critique of Practical Reason* (1788), Kant published his *Foundations of the Metaphysics of Morals,* which promised to rescue moral philosophy from vulgar prejudice on the one hand and the dogmatism of the schools on the other. It would situate morality in the precinct of pure reason. Nothing inconsistent with reason would be admitted, as had been the case with the first *Critique*. But was morality so subject to rational regulation? Kant's categorical imperative, which had many versions, is the answer: "That is, I should never act in such a way that I could not also will that my maxim should be a universal law." Thus adultery would be wrong, not because it violates the sixth commandment, but because if everyone committed it, trust, marriage, families, and eventually the human race would break down (my argument). The reasoning is secular because it is immanent; it makes no appeal to the will of a transcendent being. Nor does it completely flout Christian morality, as it arrives at the same conclusion by another route. It is individualistic, in the spirit of the Enlightenment, because man is seen (flattered?) as capable of making his own laws. He is, to invoke a Renaissance phrase, a "masterless man." Finally, it is rationalistic because, to borrow the words of a Kant title of 1793, it seeks to found morality "within the boundaries of mere reason." Kant is enthralled by reason and loath to exceed its limits. He does not acknowledge the existence of another sphere, for how would one know it?[15]

There is a check, however, to Kant's moral individualism. The maxim that the individual forms with the categorical imperative must be universally valid. Although it is subjective in its formation (the self legislates), it is objective in its result (it is valid for humanity). Furthermore, it justifies only loftiness, not hedonism. The will that forms the maxim must be pure and treat all other persons as ends in themselves, not means—one of the most impressive elements of Kantian ethics. A merchant must have in mind the need of his customer, not solely his own profit, so that he does not cheat. The professor must treat his students as human beings, not sources of tuition. The lawyer must have in mind the real interests of his client, and not see him as the source of indefinite litigation. The doctor must not invent causes for appointments or operations. Parents must not try to form the child as their spitting image. All these are cases of using people rather than treating them as ends in themselves. Kant's republic, composed of citizens possessed of pure good will, all of whom take one another as ends and not means, is a heaven on earth. It invests man with qualities one usually associates with saints or God. It divinizes man, whose will transcends the marketplace, who has nothing empirical or sensualistic in him, but who acts out of pure a priori principles rather than expediency. While his antitheological individualism fits well with the Parisian Enlightenment, as does his stress on happiness as the end of life, his emphasis on the transcendent purity of the will completely contradicts the ethics that was being developed across the Rhine on the grounds of interest, sensibility, pleasure, and utility. For Kant, a moral act must be moral in itself, not in what it produces. Happiness derives from the purity of the will, not from the acquisition of things outside it.[16]

### iii

Kant admits that, like the objects of pure reason, moral objects cannot be proven either: "To explain this, all human reason is wholly incompetent." Thus, they fall in the category of imponderables, "and all the power and work of seeking an explanation of it are wasted." We "cannot intuit its perfection."

Kant's thought soars into the ethereal and then lapses back tragically into doubt, having been rejected by the limits of reason. He is as much a post-Christian as any Enlightenment thinker, but perhaps a more sophisticated one in that he co-opts so much of Christian morality and metaphysics without being able to legitimate it ontologically. His secularism was to banish theology from philosophy, but then to welcome back elements of it that passed the test of rationality.[17]

He indeed wanted to "make room for faith," but he was far from Leibnizian optimism. In September 1791, he published "On the Miscarriage of Philosophical Tracts in Theodicy," in which he denigrated Job and his vindication of human misery in the sight of God. "Ecclesiastical" faith, with all its biblical adventitiousness, vicissitudes, and contingencies, is fundamentally "counterfeit" to the philosopher of pure, rational morality.[18]

Kant was nonetheless heard. He held forth at the table of nobles in praise of the French Revolution. After the government's publication of "Royal Edict Concerning Religion" (July 1788) and the "Edict Concerning Censorship" (December 1788), he was warned by the government not to publish anything more against religion or to "expect unpleasant measures for your continued obstinacy." Kant made the first major break between philosophy and theology on the one hand, and between morality and pure and practical reason on the other. He also juxtaposed what he considered historical, accidental dogma and pure, necessary morality.[19]

In a footnote to the second edition of his *Critique of Pure Reason* (1787), Kant added, "Metaphysics has as the proper object of its enquiries three ideas only: God, freedom and immortality—so related that the second concept, when combined with the first, should lead to the third as a necessary conclusion . . . beyond nature . . . to the highest ends of our existence."

He followed up this first *Critique* with two more, because the first did not provide an answer to the fundamental problems it raised. How was man to live a moral life if there was no God, no freedom, and no immortality? This question must seem lofty to many a modern who "muddles along," as Freud put it, either without worrying too much about morality or making do with his customs and law. Such an attitude could be described as the morality of self-interest, or of self-preservation, promoted by the English empirical school and its French imitators. But Kant did not think that any morality was possible without a belief in liberty, by which he meant free will. A paradox of the Enlightenment was that just at the moment so many were becoming inspired by the American Revolution to proclaim their own "Liberty," with a capital "L," many were denying that it derived from freedom of the will. Kant had seen their point in the first *Critique:* universal necessity was just as plausible an explanation of the world as human freedom. In the 1788 *Critique of Practical Reason,* however, he wrote, "Without transcendental freedom . . . no moral law and no accountability to it are possible." For how can you blame or praise or hold accountable a person if he is constrained to do what he does? Colloquially, we assume the existence of free will. Kant says we have to if we are to allow morality to exist. We should not praise a person who has no responsibility for an action any more than we would praise Vaucanson's robot for bowing.[20]

Kant, it seemed, felt it necessary to come up with a morality in keeping with the freedom and dignity of man, and this necessity, as we have seen, he called "the categorical imperative," of which there were many versions: "So act that the maxim of your will could always hold at the same time as a principle establishing universal law." This law preeminently presupposes freedom of the will, not only because no blame or praise is possible without it. The law is binding on the individual only because he has made it for himself. It signals the coming of the individual after centuries of "nonage," to use Kant's 1784 terminology.

Kant thought it impossible to act morally if there were no sanction to do so. This, of course, contradicts his principle of autonomy. He feared how we would be apt to act if there were no ultimate reward or punishment. If the soul is mortal and all ends at death, man can calculate his pleasures and pains as he likes (hedonism).

Without some accountability, even the motivation of making a universally valid law could be insufficient. Self-interest could not save the categorical imperative. What, then, could bring humankind into line? Only a sanction, and Kant produces two: the existence of God and the immortality of the soul. Thus, whereas in the *Fundamental Principles* he derives morality without recourse to God, here he finds God necessary as a sanction. These staple beliefs of "natural religion," eloquently defended on other grounds by Rousseau in the "Profession of Faith of a Savoyard Vicar" (*Emile,* bk. IV), were the postulates of "practical reason," the reason that comes into play when man wishes to live practically rather than speculatively.

These postulates are, of course, indefensible from the point of view of pure reason, as we have seen. They are not even exact truths in the pure sense, but fictitious norms that we must live by if we are to get on with life. They reintroduce the Averroistic double truth—one for the learned and one for the mass of men. Only savants are capable of understanding the antinomies; but every man must believe the postulates of practical reason.

Enlightenment and revolutionary men believed in Kant's postulates without learning them from Kant. Natural religion was the new orthodoxy, enshrined in the festivals of the French Revolution. Robespierre had said that if God did not exist, we would have to invent him, for "atheism is aristocratic." In the Festival of the Supreme Being on June 8, 1794, "the Immortality of the Soul" was conspicuously promoted. This occurred months after thousands of priests had been forced to abdicate their office— even abjure their faith, escape into exile, or be guillotined.[21]

### iv

In 1793, the same year as the Terror, Kant published his *Religion within the Boundaries of Mere Reason.* The title suggests something akin to French Enlightenment and Revolutionary deism, which had rejected the idea of any supernatural revelation. Kant repudiated the notion as well, but his approach was decidedly more moderate and compromising.

The main thrust of Kant's 1793 work is his critique of Revelation— something he barely mentioned previously. The Enlightenment had been very hard on Revelation. Voltaire had reviled it because of what he considered the absurdity of the Bible and its preferment of the Jews as the "chosen people." Kant was also critical, but without Voltaire's venom. Kant devised a unique historical criterion: the only legitimate and credible Revelation would be one conformable to reason. And to be so conformed, a religion had to be universal—that is, uniform in space and time. Every human being should receive the same divine enlightenment as everyone else. The absence of it poses squarely one of the gravest difficulties facing belief in Revelation (Dante was acutely aware of it). Why the Jews? Why not the Greeks, the Romans, or the Aztecs? The idea of a special revelation to one people offended Enlightenment notions of the equality of humankind.[22]

Kant's religion of reason was addressed to the enlightened, and it is one more example of the elites withdrawing from the anthropology of mankind by the end of the eighteenth century. In other words, Kant's revelation was not universal!

But was Kant prescribing how God should reveal himself? Christians understood revelation as something freely given, like grace. Was not Christianity universal with its Christ-given command, "Go, therefore, make disciples of all nations" (Matt. 28:19), and with the missions that followed?

Kant does not inveigh like Locke against a subsidized and established church, but rather bids ministers to preach faithfully their congregation's "symbol" ("On their own they can speculate"). Again, he seems concerned with guarding the moral order, rather than finding "the one true church," as was Locke. In other words, Kant is a good deal more expedient. Churches clearly belong to the realm of "practical" rather than "pure" reason. Does Kant offer a secular parody of Christianity? He recounts primitive origins without serpent, where a single couple live blissfully in a garden without pain or strife, but from which they are expelled because of conflicts over reproduction. While biblical accounts favor pride rather than lust as the cause of the Fall, Kant's story of Adam and Eve is nonetheless desupernaturalized. Evil results from man's free choice. Man is the author of his own destiny, his own salvation. Kant is a Pelagian who rejects the Christian redemption by vicarious suffering. Man saves himself. Thus, Kant is much closer to the Enlightenment's kingdom than Christ's. Miracles do not so much violate the laws of nature, as for Hume, but commit the sin of "presumptuous transcending." Humanity should be content with the pedestrian means that nature affords. There is no need to hark back to the past. This from the man whose daily walks occurred so regularly that a villager could set her watch by his passage.[23]

### v

Kant's *Religion within the Boundaries of Mere Reason* resembled his earlier (1784) "Idea for a Universal History from a Cosmopolitan Point of View." In the former, he aspired to a rational religion for mankind. In the latter, individual nations cede to a unified humanity, maintained by the regular, lawful movement of "marriages, births and death." Individual histories were "folly, childish vanity, even . . . malice and destructiveness." But reason weaves their errant, idiosyncratic strands into a larger, more coherent fabric. Religious truth is not the prize of one nation or sect, nor should peace be the affair of one generation. Again Kant strove to bring the polarities of purity and practicality together. He sought truth not in "counterfeit," "statutory" (i.e., legalistic) religions, nor in the claptrap of historical vicissitudes and bizarreries in which Voltaire flourished, but in the pure domain of the conceptual. He does not hark back to some better age. He even considers his own "the best ever"—the "visible representation of an invisible kingdom of earth." But that earthly heaven is threatened by a "hell of evils . . . annihilating all cultural progress through barbarous civilization" —a hell that haunted Edward Gibbon and Edmund Burke.[24]

Secularization has been denied outright by some contemporaries such as the sociologist Peter Berger, for religion has a way of sneaking back in. Pareto called it a "residue" of former ways of thinking—evidence of some sort of genetic survival of species-being. This in our mind is precisely what makes it secular, for secularism does not simply snuff out religion instantaneously, or at all, for that matter. The flames of belief reignite, as do inextinguishable birthday candles. In intellectual history, religion reappears in some transmogrified form. Uprooted religious ideas are grafted onto foreign foliage. Kant's secularism is one of assertive heterodoxy, individualism, and voluntarism. Human effort, for him, takes precedence over race, redemption, and miracles. He is Napoleonic in his collapsing of the distance between intention and deed. He is Roman in his confidence in human accomplishments such as the Revolution. He is oblivious to the forces within us, such as violence, that undermine and swamp our conscious intentions. His great secular wish is that, after centuries of subservience, man may throw off "nonage" and live a life of enlightened independence. That this was only conceivable, let alone feasible, for a tiny upper class did not faze him. His whole corpus is fraught with weights and counterweights of pure versus practical reason. He deprives religion of its metaphysical certitude and then restores it as a practical necessity. The dialectic, which would continue through Hegel and Marx, begins with antinomies that are not proofs but contradictions of proofs, the dialectic of transcendental criticism.

Kant never embraced either metaphysics or empiricism at the expense of the other. While his major work, the *Critique of Pure Reason,* was published in the last decade of the Old Regime, it was not until after the Terror— during the era of the Directory, Consulate, and Empire—that his works became known to most Parisian intellectuals. High-ranking politicians, such as the Directory's foreign minister Charles Delacroix and the ubiquitous Abbé Sieyès, were the first to bring Kant to the attention of the postrevolutionary brain trust. Sieyès had just left his post as minister to Berlin to become one of the directors who would usher in Napoleon.[25]

The reaction to Kant's works was mixed. Some like Delacroix believed that Kant put on paper what the French Revolution had done in practice—that is, he applied the principles of morality and natural laws to politics, making synonymous the terms "friends of philosophy" and "friends of liberty." As Tracy would put it, "I treat the Kantians with great respect because they are the truest friends of liberty and of reason in Germany." Sieyès was the most resolute proponent of Kant, whom he tried unsuccessfully to get elected to the Class of Moral and Political Sciences of the Institute. The famous philologist and pedagogue Wilhelm von Humboldt recorded in his diary of May 27, 1799, that Destutt de Tracy, the *chef de fil* of the ideologues, had organized a meeting with several ideologists, including Sieyès, where Kant's philosophy was discussed at length.[26]

For Tracy, thought was sensation tout court. There was no internal reflection à la Locke or categories à la Kant that would indicate the existence of a nonmaterial mind. This he had made clear when he coined the term "ideology" at the Institute in 1796, for, unlike "psychology," it did not presuppose the existence of a soul.[27]

Humboldt explained Kantianism as the science that fixes the limits of our knowledge that was not to be seen as passive understanding (the tendency of *idéologie*). Rather, the mind was to be conceived as an actor departing from a priori concepts rather than as the proverbial tabula rasa of Locke and the ideologues. "The French have no idea," Humboldt concluded, "of the intuition of the self outside of all experience . . . For that reason we were in two different worlds." Kant threw the well-established sensationalist psychology of the ideologues into confusion.[28]

Tracy's follower proved the first defector from ideology. Maine de Biran adopted Tracy's use of self-movement, not to prove the existence of the external world against the idealists, but to prove how a "hyper organic self" is brought into existence by the effort it exerts against the external world. Biran did not believe that sensation could be the same as thought per se, for how would it distinguish its different modes, such as judgment and memory? With that observation, Biran launched the nineteenth-century French philosophy that would veer away from eighteenth-century sensationalistic psychology and move in the direction of Kantianism and Hegelianism. His memoir, as submitted to the Berlin Academy of 1806 (along with one by Tracy), showed how far he had come to accept Kantian language and concepts.[29]

Kant's *Critique of Pure Reason* remains the great transcendental work of the century. It contributes powerfully to secular thinking in that it removed metaphysical truths from ontology, or the study of real being, to epistemology, or simply the way to perceive them. It widened the gap between the ancient and medieval cosmological focus on the one hand and the Cartesian and contemporary phenomenological one on the other. Thus, to take God: he becomes an antinomy, a possible, but not a necessary, being.

# Part IV

## Introduction:
## The Dialectic Upward and Downward

Modernity was a time of conflict for the philosophers of autonomy. How could it have been otherwise when each individual was a norm, a law unto himself? We have passed through the conflicts of the two cities, of the papacy and the empire, of politics and raison d'état, of individualism and orthodoxy, of the senses and the soul. Kant referred to some of these conflicts as the "antinomies of reason." Hegel took all of the antinomies of the past and placed them in a systematic schema, which he called the dialectic. Reality for him became eminently conflictual. The question was not one of choosing the right doctrines, but of living through their affirmation and negation as the necessary process of life. Furthermore, Hegel makes all these antinomies, which he calls antitheses, reach something better that is closer to the truth in each stage of the conflict. This process is immanent, rather than transcendent; within history, and not outside it—a secular explanation in itself. At every stage, history works toward greater self-consciousness, a greater self-knowledge that becomes equated with greater freedom. The end of history is absolute spirit, truth, and freedom. From confusion, Hegel sought to extract sense.

Marx derided Hegel's "self-consciousness," but not his dialectic. Freedom would come about when the oppressed took charge of the *economic* and *social* forces exploiting them. History was dialectical—made up of production and the resultant class conflicts, rather than of ideas.

We have passed from the city of God, and the summum bonum of the medieval era, to the "anthropology" of early modernity, which placed man at the center of all. But now many reject even that status on the grounds that "man" refers only to privileged men at the expense of others. We must now descend further to emancipate *all* men, not religiously or intellectually, as had been tried before, but physically, holistically.

CHAPTER TEN

# G. W. F. HEGEL: SECULAR PHILOSOPHY COMPREHENDING THEOLOGY

The German philosopher Friedrich Hegel had nothing but disdain for the French ideologues. Instead of grappling with man's mental faculties, his sensations and ideas, Hegel struggled titanically with the contradictions of history, which he made purposeful and rational through the dialectic. Kant's antinomies of pure reason were the origin of the dialectic, but it was Hegel who transferred them from metaphysics to history. All Hegel's philosophy, even his logic, had a historical dimension to it, just as indeed his history had a logical one. His secularization of theodicy, or the justification of the "ways of God to men," allowed Christianity to survive the French Revolution—but only by being "comprehended," which is to say, by being rationalized, or secularized. [1]

i

Hegel's father, a senior civil servant in the duchy of Württemberg, took dogmatic Christianity so earnestly that he steered young George Wilhelm Friedrich (1770–1831) to the Lutheran seminary of Tübingen. There the young man studied from 1788 to 1793 together with the poet Friedrich Hölderlin (1770–1843) and the idealist philosopher Friedrich Schelling (1775–1854). His boyhood studies included the classics, especially Greek literature, where he encountered Sophocles's *Antigone,* which foreshadowed the Christian problem of the relationship of religion and the state that became his lifetime concern.[2]

When the French revolutionary armies spilled over the Rhine in 1793, Hegel and his "Jacobin" friends allegedly danced around a liberty tree, celebrating the earth-shaking event that promised universal liberation. By 1807 the French Revolution meant for Hegel that "the human Spirit had outgrown [the past] like the shoes of a child." The ideals of the Enlightenment—liberty and rational inquiry—as well as of religious toleration clashed with those of orthodox religion. Then too, the regnant philosopher, Immanuel Kant, had recently finished his three *Critiques* and his *Religion within the Boundaries of Reason Alone*—the final Enlightenment critique of the Judaeo-Christian Revelation. This too hit Hegel hard.[3]

During a private tutorship in Berne in 1795, Hegel wrote a *Life of Jesus,* which, under Kant's rationalistic influence, portrayed the founder of Christianity as a benevolent sage rather than a miracle worker or the Son of God. Later, during a tutorship in Frankfurt in 1796, he penned an article, "The Spirit of Christianity and Its Fate." While neither of these was published during his lifetime, each shows that, early on, he was applying rationalistic principles to religion. But he did not dismiss the transcendental claims of Christianity as "fraud" or "imposture," as had the French philosophes.

Although these youthful writings of Hegel have been interpreted as politics in disguise—a revolt against despotism—a closer reading suggests otherwise. Hegel was trying to reconcile both dogmatic Christianity *and* Enlightenment rationalism. "Reason and freedom remain our password," he wrote, "and the invisible church our rallying point." The philosophes would have adopted only the first half of that sentence, but for Hegel, dogma became the basis for speculative thinking—or the reasoning that transcended dogmatics without overthrowing them.[4]

The key to Hegelian thought is again the dialectic. Two irreconcilable ideas clash and are sublated *(aufhebung)* to a higher level. In the process, the elements supposedly left behind are actually "encompassed" and included in the next stage of thought. Such antitheses—the main components of Hegel's dialectic—can be found as early as 1800. But the dialectic is not full blown until Hegel's first published and most famous work, the *Phenomenology of Spirit* (1807), which was in some ways a German counterpart to the Marquis de Condorcet's *Historical Outline of the Progress of the Human Mind* (1795), in that both showed the progress of reason through history.

Napoleon brought the French Revolution to a close in 1799, the year in which Hegel was left with an inheritance sufficient to pursue a university teaching career. Hegel was teaching in Jena when Napoleon marched through it as conqueror in 1806. The twenty-six-year-old professor recorded that he saw the passing of the world spirit. He aspired to move into a Napoleonic-controlled part of Germany (Prussia was only the emperor's ally). From that time on, Hegel saw the embodiment of the new age in the powerful, authoritative reforms that Napoleon applied throughout his annexed or vassal states. These included confiscation of ecclesiastical property and enactment of the

Napoleonic code, which embodied many of the principles of 1789, such as equality of opportunity and abolition of serfdom. Like most Europeans at this time, Hegel wanted reason, liberty, tolerance, *and* order—not anarchy or terror. Germany would adopt some of the principles of 1789 and of Napoleon, but German idealism would surpass both.[5]

After fathering an illegitimate son, Ludwig, in 1807—a son whom he reared together with his other children—Hegel in 1811 married Maria von Tucher, daughter of a Nuremberg patrician. She bore him four children in three years, during which time he became rector of the gymnasium of Nuremberg. On his later academic trips he wrote his wife frequently about museums, chiefly about his keen love of the Madonna who, he felt, honored women.[6]

Hegel's *Logic* (1812), like his *Phenomenology* (1807), was a historical and secular introduction to speculative philosophy, reaching back to ancient Greece and forward to modernity. By 1816 he was professor at Heidelberg, one of the oldest and most prestigious universities in Germany. In the next two years, the second volume of *Logic* and the first edition of his *Encyclopedia of Philosophy* were published, and the now flourishing scholar was named professor of philosophy at the recently founded University of Berlin.[7]

By 1820 Hegel had reached the apex of his career. In his only work on political theory, *The Philosophy of Right* (1821), he supported the Restoration corporate state and said, "What exists is good and rational," and that the state was the "actuality of concrete freedom." A few years later, Hegel described himself as a professor who "prides himself on having been baptized and raised a Lutheran, which he still is and shall remain." He was comfortable in his role.[8]

During the 1820s, while Hegel was teaching at Berlin, his students published transcripts of his lectures with which he was not entirely satisfied, but which have become indispensable and are more comprehensible than his own writing. These lectures were syntheses of "The History of Philosophy," "The Philosophy of History," "The Philosophy of Art, or Aesthetics," and "The Philosophy of Religion."[9]

Hegel lacked charisma and even coherent delivery in his lectures. He formulated his thoughts extemporaneously and with great effort. The obscurity of his writing led one biographer to charge him of sophism. Legend has it that Hegel complained that "only one student understood me, and he misunderstood me."[10]

But he was never so abstract as to be indifferent to appointments, audiences, protégés, rivals, and emoluments. At the end of his life, he was overshadowed by younger colleagues, who were more comprehensible. But even as late as 1829, he could attract 200 auditors to his lectures on the proofs of the existence of God. During the 1820s, he also tried to place a few dedicated students in university faculties to ensure the survival of his views. He wanted to be an educator in the Enlightenment manner, that is, "in the "criticism and reconstruction of religion." "Our knowledge of God," he wrote, "was the absolute problem of philosophy."[11]

Reared in Hellenism, Christianity, the Enlightenment, and Napoleonic imperialism, Hegel was at times plagued by charges of Spinozism, Schellingism, and atheism; he spoke of God a great deal, but it is doubtful that he believed in the traditional personal God or any substantial, existing deity. Nor was he a Romantic, being too tied to an eighteenth-century faith in reason—he detested the Romantic vogue of philology and what he considered the equally Romantic cult of "feeling," promoted by his rival at the University of Berlin, the philosopher Friedrich Schleiermacher (1768–1834). He equally deplored the lapse of so many thinkers into complete individualism, wherein each thinker developed his own singular philosophical system.

## ii

Is Hegel "the last Christian philosopher," as Karl Löwith believed? Or an atheist with a cultural fondness for the Christian "myth"? To answer is to apprehend the two ways Hegel understood religion generally. His first approach, "pictorial thinking," represents certain abstract truths or historical events, which are fraught with spiritual significance but cast in concrete images. Exodus, the Incarnation, and the Crucifixion are "pictorial" representations, regardless of their actual occurrence. In the second approach, "speculative thinking," the philosopher eschews pictorial thinking to concentrate exclusively on the abstract truth of a doctrine or event. Hegel believed that Christianity was a religion of truth, but that not all its representations were of ontological (i.e., really existing) or historical entities. They were popular expressions of higher philosophical truths, made understandable for every human. This conceptualization distinguished Hegel's secularism from that of the Enlightenment philosophes in three important ways: (1) the philosophes did not grant as great a "truth claim" to Christianity as Hegel did; (2) Hegel did not deprecate the popular or pictorial version of Christianity; and (3) he greatly attenuated the philosophes' version of Averroes's "double truth" (one for the people, the other for the savants). Instead, Hegel believed in a double expression of the same truth, one consisting of "representation," images, or *vorstellungen,* produced by pictorial thinking, and the other of abstract concepts.[12]

There are, for example, certain unique Christian doctrines such as the Trinity, which the church believes "in this childlike mode of representation as the relationship between father and son [and Holy Spirit]." But occasionally Hegel's denigration of pictorial thinking approaches atheism, such as his contention that God is an "[u]ncultivated" [*sic*] idea, or that "God's objectivity is . . . strictly only a revelation, only a manifestation of the Spirit of the Age." And again, "Creation" is a "meaningless word derived from the ordinary conception." Referring to the Cambridge Platonists, he detects a weakness in their "Christian conceptions of God and angels—all regarded as particular existent things." Angels, he repeats, are not "objective beings." Unlike Locke, who felt that miracles attested to Christ's messiahship, Hegel believes that they distract from the true message. And his thoughts on the Resurrection—the central mystery of the Christian religion, without which all preaching and faith are in vain (I Cor. 15:14)? Hegel wrote, "To consider the Resurrection of Jesus as an event is to adopt the outlook of the historian, and this has nothing to do with religion." In other words, the religious significance of the Resurrection does not depend on its historical occurrence, for it was above all a meaning rather than a fact. Hegel wanted to transcend the contradiction between the Christian affirmation and the Enlightenment denial of the Resurrection. This allowed him to accept the philosophes' historical critique of religion, yet retain its spiritual marrow. It could also be seen as a representative statement of the post-Enlightenment idealist Lutheranism, affirming the *spirit* of Christianity as opposed to Catholic corporeal objectivism.[13]

In his *Lectures on the History of Philosophy* (1822), Hegel devoted a long section to the sixteenth-century philosopher Jacob Boehme (1575–1624), revealing his ambivalence to pictorial thinking. Only recently recognized as a thinker of note in Hegel's time, Boehme was born in Upper Lusatia, where he tended cattle before becoming a cobbler. His language was sensuous, imaginative, and forceful, which led Hegel to call him (fondly) a "complete barbarian." Boehme's allusions to "qualities, spirits and angels" made Hegel's head "swim." But he appreciated Boehme's philosophical idealism: "It is the Protestant principle of placing the intellectual world within one's own mind and heart, and of experiencing and knowing and feeling in one's own self-consciousness all that formerly was conceived as Beyond." Indeed, Boehme anticipated Hegel's theology in certain respects. The cobbler's Trinity, for instance, derived from "God's diremption [self-division] of Himself;" from which also came Lucifer, or evil. Boehme had "the profoundest idea of God, which seeks to bring the most absolute opposites into unity." He represented "not . . . the empty unity, but . . . this self-separating unity of opposites." Like Hegel's thinking, Boehme's was dialectical, with pantheistic overtones. "The world," according to Boehme, "is none other than the essence of God made creaturely."[14]

Hegel's "speculative" versus pictorial thinking is somewhat akin to Kant's distinction between "pure" and "practical" reason—the first is the language of the philosoper, the second that of humankind; the first is abstract, the second filled with images. For Hegel, the primary object of speculative thinking is to know God, even though such thinking may not affirm his existence. The very inadequacies of pictorial thinking presuppose a need for speculative thinking. "If . . . it is assumed as known what God is, how He is . . . then one could not at all imagine what philosophizing was still for, because philosophy can have no ultimate purpose other than to perceive God."[15]

To state that speculative thinking was invariably irreligious or atheistic would therefore clearly be a mistake. Indeed, confronted with certain passages suggesting the divinity of Christ, Hegel admonished, "The question is not whether exegesis [can] of itself flatten out these expressions." "Only recent philosophy has attained this conceptual depth" of "truth in and for itself."[16]

Speculative thinking is best explained as an activity of the mind that stretches to the outer reaches of being. It is the very opposite of the British and French empirical tradition, which strives to anchor man's thought in the sensuous immediate lest it lose itself in abstraction. For ideology, the revolutionary representative of this empirical school, God and soul are definitely outside philosophical boundaries.

Hegel's philosophical pretensions knew no limits and distinguished his philosophy from that of Aquinas and Kant, both of whom spoke of limits to reason. For the Scholastic, philosophy was limited by Revelation and called the "handmaid of theology." In that Christian tradition, philosophy could approach the outer gates of theology, but stood in awe outside the inner sanctum. For Hegel, philosophy was no handmaid, but mistress of the sciences.[17]

\* \* \*

To the uninitiated, Hegel may seem a very religious philosopher, since the words "Spirit" and "God" appear constantly in his writing. Yet some critics, most notably Alexandre Kojève, considered him an atheist. Hegel, however, was no Enlightenment scoffer. As Clark Butler puts it, "Speculative philosophy preserves faith in raising it to the level of knowledge." The question, of course, is whether by "raising" faith to the level of reason, Butler assumes that knowledge is greater than faith, which then would lose its supernatural character. Hegel's secularism did not so much exclude religion as deprive it of its transcendence.[18]

Hegel's God is certainly grander than the Enlightenment watchmaker's, for he must be sought in the outer limits of consciousness rather than in mechanical philosophy. Hegel is a secularist in that he does not accept, as did Aquinas, the limits imposed on reason by Christian Revelation. But he is a novel secularist in considering Revelation at all.

In his *Lectures on the History of Philosophy,* Hegel throws out most of the medieval, Thomistic proofs of the existence of God, virtually without examining them. One, however, caught his attention, St. Anselm's so-called ontological proof. Hegel considered it brilliant in that it made God's existence proceed from his definition, that is, from the very concept of infinite perfection. (To be infinitely perfect necessarily entailed existence, for if it did not, God would not be infinitely perfect.) It is not hard to see why this definition appealed to Hegel: it required no external, contingent, or empirical observation or fact as did, for instance, Aristotle's argument from motion. It began in thought and ended in thought; it was proof through the very idea alone—idealistic in the sense of Kant, Fichte, and Schelling. Ironically, Hegel half-heartedly conceded Kant's antinomy on God's existence on the grounds that it unjustifiably passed from the logical order to the existential order and from the finite order to the infinite order.[19]

Hegel believed that proofs of the existence of God were in dissolution in Restoration Germany. As early as 1807, in the *Phenomenology of Spirit,* which established his reputation as a philosopher, he broached the subject as follows:

> The Lord and master of the world holds himself in this way to be the absolute person, at the same time embracing within himself the whole of existence, the person for whom there exists no superior Spirit. He is a person, but the solitary person who stands over against all the rest. These constitute the real authoritative universality of that person; for the single individual as such is true only as a universal multiplicity of single individuals. Cut off from this multiplicity, the solitary self is, in fact, an unreal impotent self.[20]

The "Lord and master," the "absolute person," the "solitary person" greater than whom no one is—these descriptions of God are completely compatible with traditional theology. But Hegel dialectically affixes the universality of all other individuals, without whom, he avers, God is impotent. Hegel augments the importance of humankind in contrast to Christian and Jewish theologians who underscored the gulf between creature and Creator.

For Hegel, humanity is indispensable to God. Through the dialectic, man becomes more and more self-conscious, purer, and closer to Absolute Spirit, absolute self-consciousness. Man even becomes God or part of God. "God's objectivity," Hegel wrote, "[is] realized in the whole of Humanity." This objectification of God in the human spirit takes place in time; hence the importance of history.[21]

Secular knowledge, Hegel felt, was outstripping religious knowledge, and he wished to conflate the two by justifying the ways of God to man through philosophy and philosophical history. Still he could not philosophize in the manner of a Eusebius or a Bossuet, that is, as a theological historian. Nor could he accept wholly the Enlightenment's secular attitude toward theology. The Enlightenment, he thought, "has sinned against the Holy Spirit." "Unhappy the age that must content itself with being forever told that there may be a God."[22]

Yet Hegel's idealism is as responsible as anyone's for this uncertainty about God's existence. His *Aesthetics* includes this remarkable sentence: "We must regard the existence of the Spirit *in the consciousness of man* as the essential Spiritual existence of God" (emphasis mine). If God can be seen as existing *essentially* in the thought of humankind, does he have an existence outside human thought? The question, Hegel explains, is, "How by means of thought God is to be brought about?" Is God, then, "brought about" by man, rather than man by God? In the *Lectures on the Philosophy of Religion* he elucidates: "God and religion exist in and through thought and solely in and for thought." Such a statement can be read as an expression of subjectivist idealism or as atheism, or both, for the first lends itself to the second. Hegel leaves no doubt about this when he says that God is "a fact of my consciousness, not objectively true." In sure anticipation of the atheism of Ludwig Feuerbach (1804–1872), Hegel writes unambiguously in the *Phenomenology:* "Conscience is certain of itself . . . which, when posited outside the self is an abstract God and hidden from it—such that the difference is eliminated."[23]

In the mid-1820s, Hegel speaks critically of the "religious sensibility of our time," which consists in "one's consciousness of the divine as self-consciousness, *oneself* as divine—deification of oneself," but he contributes to this sensibility. "By consciousness of the unity of divine and human nature," he writes elsewhere, "we mean that humanity implicitly bears within itself the *divine idea* . . . as its own substantial nature" (emphasis his).[24]

When early Christians spoke of man becoming like God, they meant that by divine grace they could participate in divine life. They did not, as did the serpent in Genesis, think of man ontologically becoming like gods. For Hegel, the Incarnation was the quintessence of this union of humanity and divinity, which represented "the consummation of reality . . . the most beautiful point of the Christian religion." If one were to deny all the miracles of the Gospels, he maintained, there would still be the miracle of Christ's character—a sentiment he shared with Locke and Rousseau. The only question Hegel has is whether God becomes man in a unique moment of history, or whether at that moment man becomes conscious of his own divinity. If it is the latter, there is no limit to man's being, to his finitude. Hegel does not choose.[25]

In his *Lectures on the Philosophy of Religion* (1827), Hegel firmly disassociated himself from *pantheism*. Unlike Fichte and Schelling, Hegel *can* be regarded as a dualist in that he saw everything as a struggle between subject and object, both of which are presumed to exist. He emphatically retained the Kantian bipolarism (making it dynamic), which Fichte's and Schelling's idealistic reductionism had largely eliminated.[26]

But Hegel's reader is always there to nettle him with the question, Is God more than a thought? Indeed, a follower of Hegel, Philip Konrad Marheineke (1780–1846), aspired to "a definition of religion that affirms God's actual existence, not merely the thought of God." Hegel broached a definition in his *Lectures on the Philosophy of Religion* (1824). He said that "God as spirit, who is infinite subjectivity, infinite determinacy within himself, is both the absolute truth and the absolute goal of the will." Does that unequivocally make God more than just the subject's thought?[27]

Hegel accepted the conclusions of the Enlightenment without apparently rejecting Christianity; he eluded the hard choice between them thanks to the dialectic, which transcended their contradictions. His world was immanent rather than transcendent, for he wished to substitute Spirit for an ontological God. He was a gnostic in the sense that, for him, the highest reality was the knowledge of Spirit divorced from the corporeality of the incarnate God-man, which was, for him, more idea than flesh. He showed how the purely spiritual could be preferred to the incarnate God, before Marx rejected this purely spiritual God for the purely material "species being".[28]

### iii

Hegel did not simply reach a standoff with religion as Kant had in his antinomies of pure reason. Hegel's thinking was far too dynamic. He wished to "encompass"[29] a superannuated religion, which could survive under the aegis of reason's new shoot—"transcendental philosophy." The tack he took was historical: only by the dialectic's sweep through history could truth be discovered. Logic itself was a result of historical experience.

In his 1824 *Lectures on the Philosophy of Religion,* Hegel lashed out against contemporary attitudes toward the relationship of the two disciplines of theology and philosophy. "Rational theology" was the heir of the Enlightenment, particularly Kant's Enlightenment, which denied that philosophy could prove anything concerning God. Hegel, on the other hand, believed that "God rules the world as reason" and that "philosophy can have no ultimate purpose other than to perceive God."[30]

Speculative philosophy, unlike medieval philosophy, did not proceed *from* the assumed truth of the Christian faith, but rather moved *toward* that faith. Its origin, Hegel asserted, was in reason alone, and he did not stop at the threshold of Revelation but sought to penetrate well inside it. Peace between the two, he continued, would not come through a division of labor, as Aquinas had hoped, but through philosophy's appropriation of "the long existing Christian faith." Speculative theology "preserves faith in raising it to the level of knowledge," that is, to the point where it ceases to be faith. Then "religion and philosophy coincide in one."[31]

This solution to the conflict between theology and philosophy is noticeable in Hegel's early work. Going right to the source of Western philosophy, Hegel asks, "Is Judaea then the Teutons' Fatherland?" Hegel's answer can only be no—for the Jews, he argues, have an external religion of obligation, which ignores the interiority of the spirit. The Jews lack that all-important Hegelian quality: self-consciousness. They have, he claimed rather, a "completely spiritless universality."[32]

Socrates, for Hegel, was the greatest exemplar of the philosophic mind. He investigated the truth freely, leading others to it by dialogue. His disciples were at complete liberty to let their reason ramble. They were not told what to believe, as were the Christians; nor were his beliefs and morals "given." The Greek spirit was imaginative and sensuous. Its religion was anthropomorphic and therefore congenial to man. It canceled the God-man antithesis by making God in the image of man and by making man comfortable with himself. How this Greek philosophy differed from the religion of Jesus! Hegel represented the apogee of this man-centeredness of Western humanism, which nourished itself on Greek anthropomorphism. By contrast, Christian "positivity" stemmed from its authoritarian character derived from the Jewish imperative, and Christianity spread by preaching and proselytizing rather than by Socratic interrogation.[33]

What Socrates figured out from within became something external and "given" in Christianity. Christ told his disciples to "go out and preach," and that only those who accepted his teachings would be saved. The later positive Christianity stripped primitive Christianity of much of its appeal; the practice of holding property in common, mentioned in the Acts of the Apostles (2:44–45), ceased, and then, Hegel argues, an ecclesiastical hierarchy supplanted the original apostolic equality. Later still, when Christianity was adopted by the Roman Empire, man lost his religious autonomy.

In *The Life of Jesus* (1795) Hegel showed his respect and admiration for Jesus, in whose mouth he put rationalistic discourse but whose miracles, resurrection, and claims to divinity he passed over. Hegel's Jesus was no God, but a Christianized Socrates, just as his theology was a philosophized theology.[34]

More than Locke and Rousseau, but like most "enlightened" contemporaries, Hegel felt that reason had to brush away the "cobwebs of . . . incomprehensible . . . dogma"—as well as miracles, Providence, Revelation, and even a personal God. He shunned "monkish cowardice of thought" and all the delusions of "positivity," such as images and painting. He also noted that the worldly element (for example, ecclesiastical property) in the church constituted a clerical secularism, as Marsilius of Padua had first observed in the early fourteenth century. The church's *plenitudo potestatis* under Innocent III and Boniface VIII set a mundane tone for the church, against which the Reformation rebelled.[35]

Thomas Aquinas is the preeminent master of the problem of reason and faith in Christianity, as we have seen; yet Hegel devoted only half a page of his *Lectures on the History of Philosophy* to him! No mention of the theologian's famed distinction between reason and faith, particularly relevant to Hegelianism. It is virtually certain that Hegel never read Aquinas. Hegel emphasized the nullity of medieval philosophy, barely mentioning Augustine either. It is quite likely that his Protestant hostility to the Catholic clergy (he once referred to the Jesuits as "vermin") explains his cavalier dismissal of medieval philosophy as nothing but religion lacking an antithesis, which was the precondition for truth. Summing up his disgust, he said, "This Scholasticism, on the whole, is a barbarous philosophy of the finite understanding without real content, which awakens no real interest in us . . . Scholastic philosophy is this utter confusion of the barren understanding in the rugged North German nature."[36]

There is another paradox of Hegel's view of the Middle Ages. His critique of scholasticism boiled down to the charge that it suffered from an excess of religion. But his critique of medieval civilization was that it suffered from an "unregulated . . . externality" (or positivity), by which he meant the hierarchical, institutional, and authoritarian church and its external cult of piety. Only someone like Hegel, steeped in Lutheranism, could prefer Germanic to Roman externalism.[37]

Hegel traced the origin of what he called the "unhappy consciousness" back to ancient Judaism. (The Lutherans, he averred, freed themselves from Catholic externality after a long battle with the Spirit.) What is puzzling is that Hegel, who laid such stress on the objectification of freedom in his *Philosophy of Right,* could have spurned the objectification of Christianity in the Middle Ages. If the nineteenth-century Prussian state could be thought to actualize freedom, could not the Carolingian or papal states be seen to reify freedom sui generis?[38]

Hegel thought that the most important mark of his philosophy was the substitution of "subjectivity" for medieval externality. "All externality in relation to me is thereby banished, just as is the externality of the Host." "The enjoyment of life" is not to be sacrificed; "monkish renunciation is renounced." Above all, "freedom of spirit is planted in embryo." *"Divinity is brought within man's actuality"* (emphasis mine). Man began to be reconciled "with himself." The unhappy spirit of positivity, which began with Judaism and continued in medieval Christianity, was overcome at last.[39]

Clearly, Hegel considered the Protestant Reformation a liberating movement, which was crucial to freedom and indispensable to his religiosity. It enabled man to overcome alienation—the surrender of his spirit to external sources—and to "come into his own," a phrase crucial to all nineteenth-century Hegelians.

Compared with secularism, the Lutheran Reformation was, as suggested, paradoxical. Lutheranism was more religious in that it was more "inward." The Bible-toting pietist was less secular than his Gallican or Josephist counterpart in the eighteenth century, but Lutheran "inwardness" played a role in the genesis of modern German secular philosophy. "Our universities and schools are our church," Luther vaunted. "The general intellectual and moral culture is what is holy to Protestants." One could not ask for a clearer statement of the "transfer of sacrality" from the clerical sphere to the Protestant public cultural sphere. Protestantism in general was a watershed in the history of secularization.[40]

Clearly, too, the sacralization of worldly "callings" (employments), as well as of marriage and the family, gave the laity a more honored status in the Protestant world. In the economic sphere, the Napoleonic secularization and redistribution of clerical lands among the laity tipped the balance in its favor. Finally, the multiplication of Protestant sects had the effect, observed by John Locke, of making "everyone orthodox to himself"; no religious uniformity was possible in the modern world.

But, ironically for Hegel, the Protestant Reformation was secular not because it was sectarian, but because it enabled humans to discover themselves through the subjectivity of the Spirit. The Reformation, for Hegel, was not theocentric, but homocentric, a religion that brought out what he considered the best in man—freedom and self-consciousness, the two goals of the dialectic.

Although a "Jacobin" when he authored his "Tübingen Essay," he denigrated Enlightenment snobbery and high-sounding phrases such as "the history of mankind," "happiness," "perfection"—"phrases of babbling quacks peddling shopworn panaceas," and "idle chatter," nothing but a "bellyful of book learning" devoid of life.[41]

In the *Phenomenology of Spirit,* however, Hegel praises the Enlightenment for its audacity in penetrating the furthermost recesses of religion and human nature. Still, in typically dialectical fashion (finding the seeds of destruction in every truth that led to a higher truth), Hegel also found in the Enlightenment a "sin against the Holy Spirit" (doubtless its desire to crush Christianity). *There* was the antithesis—that is, the Reformation's reaction against a medieval excess of religiosity and externality, and the Enlightenment's reaction against the fanaticism of the religious wars, which led to a later excess of impiety and ego in the French Revolution. Hegel aspired to bring a final "peace between the final religion and the final philosophy."[42]

That philosophy, unlike the Enlightenment's, would be profoundly historical. Even Hegel's logic unfolded in "historical moments": "The process displayed in History is only the manifestation of Religion in Human Reason—the production of the religious principle which dwells in the heart of man under the form of Secular Freedom." The Enlightenment lacked a true reification, or objectification, of religion. It championed light over obscurantism, reason over superstition, freedom over bondage, truth over error, and proof over prejudice, but it did not actualize these ideas.[43]

Every statement in Hegel's *Phenomenology* contains within itself its own nemesis, the seeds of its own destruction. This "ceaseless weaving of the Spirit," its inwardness, is "the defilement of Enlightenment through the adoption . . . of a negative attitude, that is, an object for faith, which therefore comes to know it as falsehood, unreason and as ill-intentioned, just as Enlightenment regards faith as error and prejudice." In other words, the Enlightenment that censors faith as error and prejudice is itself censored for its bad faith. "Thus [what] Enlightenment declares to be an error and a fiction is the very thing that Enlightenment itself is." Its "unhappy consciousness" consists of its complacency, which is based on "a mean conception of life." The Enlightenment "remains satisfied in this world."[44]

Hegel treats the Revolution in the same chapter of the *Phenomenology* ("Culture") as he does the Enlightenment, viewing it as completing "the alienation of Spirit." For a while Hegel thought that the French Revolution was a restoration of his beloved Athens. In the *Phenomenology* it is seen as an extreme development of human freedom, which, having no limits, became self-destructive. In his *Philosophy of History,* Hegel wrote, "In the French Revolution, therefore, the republican constitution never actually became a Democracy: Tyranny, Despotism, raised its voice under the mask of Freedom and Equality."[45]

This is the moment when "the owl of Minerva rises to flight with the coming of dusk." Christian civilization had to die before it could be understood. Deceased, the supernatural had left it. All that remained was morality, now the true (human) divinity. (How much did not Nietzsche owe Hegel?)[46]

Hegel's secularism culminated in the sublation of all individual, religious, and political life in the state, just as his history encompassed the past. In his *Philosophy of Right,* he showed how all forms of personal, familial, civic, and political life are so many successive actualizations of freedom, which climax in the supreme objectification of the state.

This apotheosis of the state began as the German reaction to the French Revolution. Throughout the Napoleonic period, including the patriotic years of the *freiheitskrieg,* which finally roused the German "nation" to expel the French armies in late 1813 and early 1814, he saw the World Spirit passing from France to Germany: it was not yet the revolutionary Parisian spark that ignited much of Europe in 1830 and 1848, but rather that of the German state under Karl von Stein (1757–1831) after the battle of Jena (1806). This was not an endorsement of authoritarianism, but rather a sublation of all lower freedoms and forms of collective life in a supreme objectification. The state is actuality, and "freedom is the world actually."[47]

Hegel believed, like Locke and Rousseau, that the church should be subordinate to the state. The cost was the loss of the traditional distinction between their separate spheres of activity, which had guaranteed the church a measure of liberty. Subordination not only increased the power of the state over the church and citizens generally, but accentuated its secular character as well. The state, not the church, was now the highest ethical body—indeed, a secular church. Writing in his 1824 *Lectures on the Philosophy of Religion,* Hegel stated,

> But since the secular world also constitutes human freedom, it resists this demand [the hegemony of the Catholic church in the secular world] fiercely. So the true realization of religion in the worldly sphere is the inward realization, namely, that a just and ethical civil life should be instituted. But, inasmuch as such a civic life is now established, it absorbs all of this expansion of the divine, being itself divinity in this field.[48]

In the same work, he wrote, "The state is the march of God in the world; that is what the state is." But he also wrote that, ideally, the "state as a civil state should have no faith at all." He repeated the sentiment in the *Philosophy of Right,* saying that the state was an exclusively *human* property, as were knowledge, science, and government. He did not want the church to interfere in any way with the state, which had to fulfill many of the church's former responsibilities. But this did not mean that the church had no right to exist in the state; rather, it was "an integrating factor in the state, implanting a sense of unity in the depth of men's minds, [such that] the state should even require its citizens to belong to a church." Like Locke, although less firmly, Hegel seemed to exclude atheists from citizenship. But he saw the need for the state to tolerate different religions. Lutheranism evoked deep interiority, thought, reading, schooling, and knowledge. Unlike the French terrorists, Hegel preferred to encompass Christianity rather than abolish it. Like the French revolutionaries, he espoused Jewish emancipation on the grounds of common humanity.[49]

### iv

Is God then transcendent or immanent? For Hegel he had become Absolute Consciousness. He believed that God was immanent—a kind of collective human consciousness, which, absolutized, could be called God. But God, according to Hegel, was not alien to modern man as he thought he had been to Judaic and medieval man—the "other" over and above the self. For "God himself lies dead."[50]

By the very act of God's death for man, man has become divine. But Hegel draws back from the full consequences Nietzsche would draw. Hegel does not want to do away with Christianity the way many of the eighteenth-century philosophes did. The originality of his secularism is that it could "liberate" man from God by seeing God as a product of man's thought, not as someone alien to him. He falls short of outward atheism. He continues to speak of God with respect. Nor does he wish to crush "the infamous thing," the church. On the contrary, he treasures religion and Christianity, wishing to "encompass" them rather than to demolish them. He is a sophisticated secularist—like the atheist art historian who loves the sculpture of Chartres and listens to Handel's "Messiah" in December.

But does God exist? Hegel's answer is yes and no. His is a more tortured and dynamic contradiction than Kant's. It sums up the compromise, the modus vivendi, for Christianity in the Restoration (for restorations never restore everything). The owl of Minerva had taken flight. The consequences were yet to come.

# Chapter Eleven
## Marx: "The Christian State," "The Jewish Question," and "Species-Being"

Marx, a Jewish-born Hegelian, linked the dialectic to historical materialism and posthumously provided the Bolsheviks with a powerful weapon to overthrow Russian czarism and enthrone Soviet communism in its place. Marxism became by far the most important belief system since Christianity. As an ideology, it eventually engulfed over a quarter of the globe. Its relationship to Judaism was complex. Marx was against all religion. He was also interested in the "emancipation" of all peoples and was an enemy of the Prussian Christian state, which discriminated against Jews. But the price he would exact for Jewish emancipation was very high. *On the Jewish Question* (1843) foresees a communist society in which neither Jew nor Christian would exist. Needless to say, the end result was a thoroughly secular, even atheistic, society, unless Marxism itself could be termed a religion. However, the honeymoon of the Jews with Bolshevism proved to be short. In the years that followed, the Jews began to be persecuted and were prohibited from emigrating. The question remains, Did Marx see Jewish emancipation as an obstacle to thoroughgoing secularization and to the creation of one "species-being"?

### i

The course of Jewish emancipation, from the French Revolution to Marx's *On the Jewish Question* (1843), presaged the decline of Christianity as *the* religion of Europe's public sphere, as well as the assimilation into European society of many European Jews who eschewed their Judaic religion.

The Jews had been marginalized in European society since the diaspora of A.D. 70. Legally restricted to ghettos, forbidden to own real estate, and excluded from guilds, liberal professions, and public service, they were deprived of most civic rights, including suffrage. Medieval injunctions against

Christian usury relegated this hated practice to Jews, who consequently served as the bankers of rich and poor Gentiles alike (public banks hardly existed), repossessing the property of those who defaulted on their loans. When resentment peaked, the Jews were expelled and their fortunes seized.

The Jewish ghetto sprang from the "Christian state." The New Testament blamed the Jews for Christ's crucifixion, which was no ordinary one, for Christ had admitted to being the "Son of God." Those who opposed Christ and his followers came to be known in the New Testament as "Jews," and were recorded therein as saying, "Let the punishment for [Christ's] death fall on us and on our children!" (Matt. 27:25). For these reasons, the Jews were abhorred. The persistence of bigotry had grassroots reasons, both theological and economic. Moreover, the distinct physical demeanor of the Jews (long beards, swarthy complexions), their degrading trades, such as peddling old clothes, and the observance of strange religious rites, often at variance with local law, all led Europeans to view them as foreigners. Even the Enlightenment, for all its advocacy of tolerance and brotherhood, fell far short of advocating their toleration or emancipation. For one, the philosophes abjured their "invention" of revealed religion and resented their commercial or financial dealings. Voltaire cried to high heaven more than once about being swindled by a Jewish book dealer. But over the years, a few did take up the Jewish cause.[1]

On the eve of the French Revoltion, one of the leading exponents of Jewish emancipation was a Jansenist abbé from Lorraine, Henri Grégoire, who entered an essay competition of the Royal Academy of Metz in 1785 to answer the question, "Are there means to make the Jews more useful and happier in France?" The title of Grégoire's response, *Essay on the Physical, Moral and Political Regeneration of the Jews* (1789), revealed a complex attitude. The Jews should have equal rights, he wrote, but they did not yet possess equal human attributes. These had to be "improved"—a thought already suggested by a German author, Christian Wilhelm von Dohm, in 1783. Grégoire seemed to accept some of the prejudices of the Alsatian population against the Jews that were based on usury, the use of the Talmud, and their tendency to form "a nation within a nation" (obviously encouraged by the very existence of the ghetto). But unlike other critics of the Jews, Grégoire shifted the onus of responsibility to the Christians: "Nations avow while groaning that this is your work! The Jews produced the effects, you have created the causes: who are the most culpable?" Had the Jews not been ghettoized, he implied, their natures would be different. "Christians seeing in Jews the authors of a deicide, forgot sometimes the example of their founder, who prayed on the cross for his executioners."[2]

The significance of Grégoire's essay, and of his later intervention on behalf of Jewish emancipation in the Revolutionary Constituent Assembly, was magnified because he was a priest, and the impetus for excluding the Jews from Christian society came partly from priests such as the deputy abbé Jean Siffrein Maury (1746–1817), or bishops such as La Fare, bishop of Nancy.[3] But Gregoire's efforts were paralleled, and perhaps even surpassed, by the leading orator and parliamentarian of the Constituent Assembly, Honoré Gabriel Riquetti de Mirabeau, who aggressively supported Jewish civil rights and emancipation. Seeking the universal regeneration of mankind, many members of the Constituent Assembly soon realized that their Declaration of the Rights of Man and Citizen (1789) could not be restricted to Catholic Frenchmen but had to apply to Jews, Protestants, actors, mulattoes, and eventually slaves. Emancipation would come in stages—Protestants, actors, and Sephardic Jews in 1789, all other Jews in September 1791, mulattoes in 1791, and slaves in 1794.[4]

Grégoire's enlightened humanitarianism made Christianity, in this instance, an ally of the Enlightenment and the Revolution. But some feared that it would compromise pure Christianity. To understand where the French Assembly stood witness, another cleric deputy, the Benedictine Dom Gerle (1736–1805), proposed in April 1790 that Catholicism be declared the religion of state. The motion was defeated. This meant that France no longer could be defined as a religious entity, the "eldest daughter of the Church," a title it had enjoyed since Clovis's conversion in the fifth century. If France were truly secular, it could no longer employ religious reasons for excluding the Jews from citizenship. The Jews were emancipated a year later.[5]

The decree of September 27, 1791, liberated the Jews from all previous disabilities (and "privileges") and required them to take a civil oath and assume the same obligations as other Frenchmen.[6]

The Napoleonic experience was not all positive. The emperor had formed a negative impression of the Jews in eastern France, and when he returned to Strasbourg from the battle of Austerlitz, he heard many complaints, such as the outlandish charge that half the land in Alsace belonged to the Jews. Thus, in 1807, he recreated the ancient Sanhedrin to regulate Jewish affairs. Decrees, promulgated on March 17, 1808, included the "infamous decree" against usury, which canceled a large amount of Jewish credit. The Jews were henceforth obliged to perform military service without being able to use a substitute, as could other citizens. Consistories, composed of laymen and rabbis, resulted in a "centralized hierarchy and police surveillance." The Jews greatly resented the Napoleonic "organization" of

Jewry, which was corporative rather than egalitarian, and punitive rather than liberating, unlike the 1791 decree of emancipation. Nonetheless, most Jews through the century saw the whole revolutionary epoch (1787–1815) as glorious compared with earlier and subsequent periods. As historian Maurice Bloch noted, a new Trinity (Liberty, Equality, Fraternity) had replaced the Christian one. But, at the bottom, it was the philosophes who initiated the secularization of Christianity. The Abbé Grégoire saw no incompatibility between Christianity and toleration of the Jews, but rather a rightful consequence. Resistance and resentment had not completely subsided, however. One of the leading Catholic traditionalists of the period opined, "Jews cannot be the same, whatever one does; [they] will never be citizens under Christianity without being Christians." It would follow that if the Jews *were* to be citizens (and of course, they already were), it could only be in a non-Christian society.[7]

The nineteenth-century Left was often as opposed to the Jews for economic reasons as the Right was for religious ones. Charles Fourier (1768–1830), for example, considered them usurious, "unproductive and deceitful," and favored their removal from France to a separate state, which would be created by a Rothschild. The root of socialist anti-Semitism arose from its aversion to interest on money that had disastrous effects on peasants and workers. Socialists were living in an age of finance and of industrial capitalism, both of which they condemned along with Jewish capitalists.

At the turn of the nineteenth century, the situation of the Jews in Germany may have actually been better than in France, with some notable exceptions. As in France, the Jews in Germany had historically been limited to certain trades and settlements. Frankfurt was the leading financial center of the "country," and consequently many Jews lived in its ghetto. But the dispersion of political authority also meant that the policy toward Jews was not uniform throughout Germany.[8]

The German Enlightenment had a mostly positive effect on Jewish emancipation. Moses Mendelssohn was the most important thinker of the Haskalah, or Jewish Enlightenment. He stood as an intermediary between the Jewish community and the Gentile Enlightenment. Mendelssohn worked for the edification of his fellow Jews, for greater control over their own affairs, and for the elimination of professional exclusions. In many Berlin salons of the century the Jews began to mix with the Christians, while some became members of the Masonic lodges, which were multiplying. Mendelssohn was also a frequent and respected correspondent of Kant. For Mendelssohn, assimilation may have been more important than secularization.[9]

Christian Dohm wrote about Jewish concerns as an outsider. His *Uber die burgerliche Verbesserung der Juden* (1783) envisaged the separation of church and state as the centerpiece of emancipation, for if the state were not upholding Christianity as the official religion, there would be no religious reason to exclude the Jews. Secularization, for Dohm, was more than a precondition to emancipation, as it was for Mendelssohn. Accepting full civic responsibilities, he felt, necessarily entails dispensing with conflicting (e.g., dietary) obligations of Jewish law.

But did the Jews really want secular assimilation under these conditions? Did the Gentiles really want brotherhood with the Jews, who were so different from them? How far could fraternity stretch before snapping?[10]

Forces working for the assimilation of the Jews in Germany included the schools, mostly in the northwest and Rhinish Germany, that began to admit Jewish students at the end of the eighteenth century. States like Berg and Westphalia were affected by Napoleonic reforms, which included Jewish emancipation. Prussia did not rush to act in a similar vein until after its defeat by Napoleon in 1806, when emancipation came as part of the reform by Stein and Hardenburg. Their edict of 1812 conferred "the same civic rights and liberties as those enjoyed by Christians," with the exception of employment in nonmilitary governmental posts. "This was a remarkable law," wrote historian Reinhart Rürup, "which to this day must be valued as one of the great documents of the history of emancipation." Legal emancipation, however, suffered a setback during the Metternichian era, but was completed in the 1860s under the chancellorship of Otto von Bismarck. Important goals were met, such as the abolition of religious tests for admission to universities and the emergence of some Jews in politics. But anti-Semitism resurfaced during the two-decades-long economic depression of 1871, and it reached a hysterical level in the Dreyfus Affair of 1894–1906. The social and political laws concerning Jews in Europe between the French Revolution and 1848 do much to explain the rise of the single most important secular movement of modern times, communism.[11]

## ii

Born in 1818 of Jewish parents and a descendent of a long line of rabbis, Karl Marx was baptized a Christian at age six. His father, Hirschel Marx, a lawyer in Rhineland Trier, became a Lutheran Christian, allegedly to free himself of the disabilities that most German governments imposed on the Jews who sought to enter the liberal professions during the Metternichian era (1815–1848). In 1836, Marx began studying philosophy in Berlin, where Hegel's influence was still very strong. David Strauss, one of the Young Hegelians, as they were called, had recently published the devastating *Life of Jesus* (1835), which shocked Germany by stating that the Gospel accounts of Christ's life could not be taken as history, but were only ancient Jewish myths retold. Indeed, Christ might not even have existed. Previous rationalist biblical criticism, such as that of Reimarus or P. Bayle, had contested certain historical facts. But Strauss's book believed the life of Jesus to be wholly explicable in terms of Jewish culture.[12]

The Hegelian philosophical legacy in Berlin determined the language Marx used to express his revolutionary philosophy, for Hegelianism was very much alive in Berlin five years after the death of its towering creator. How did Marx, the sole Jew by birth among Hegel's prominent followers, use Hegelianism to respond to the "Jewish question"? And did that response help shape modern secularism?

Among those reconciling Hegel with Christian orthodoxy were the "Hegelian Right"—Karl Rosenkranz, Hegel's official biographer, Eduard Gans, and Konrad Marheineke. They believed in the objective existence of God, the immortality of the soul, and the divinity of Christ. Those constituting the "Hegelian Left," also known as the Young Hegelians, were Ludwig Feuerbach, Bruno Bauer, Max Stirner, David Strauss, and the eighteen-year-old Karl Marx, who together steered Hegelianism in an atheist direction.[13]

The Hegelian Left began with religion—where it would remain for a decade—until Marx began to write. In 1841, Marx completed a doctoral thesis on the Greeks Epicurus and Democritus, whose materialism he imbibed full-heartedly. In 1843, he criticized Hegel's *Philosophy of Right* as being too taken up with "self-consciousness," divine will, freedom, personality, and other "abstractions."[14]

Two other Left Hegelians, Ludwig Feuerbach (1804–1872) and Bruno Bauer (1809–1882), contributed to the development of Marx's atheistic materialism and his rejection of Christian transcendentalism. In 1841, Feuerbach published his *Essence of Christianity,* no less startling than Strauss's *Life of Jesus.* Feuerbach's work laid forth a bold thesis that was derivative of Hegel's immanentism. God and all other supernatural beings, said Feuerbach, were creations of men's minds, projections of their own beliefs about perfection. Everything that men were in idealized form became God. These projections, originally part of man, were "alienated" into powers over, against, and apart from him. Theology in this school was really "anthropology." Its subject was man and not God. "The nature of faith, the nature of God, is itself nothing else than the nature of man placed out of man, conceived as external to man," Feuerbach averred. Strauss had made Jesus a myth. Feuerbach made Christian belief an illusion. Recall that in 1827 Marheineke begged Hegel to make "a definition that affirms God's actual existence not merely the thought of God." But Hegel never did decide clearly for the "actual existence of God"; rather, he took God as just an idea, and that was a source of the problem. His students were bolder. Feuerbach thought that he was being entirely consistent with the master: "On this process of projecting self outward rests also the Hegelian speculative doctrine, according to which *man's* consciousness of God is the *self-consciousness* of God." Man is then emancipated from his self-created taskmaster, whom Freud would interpret as the "super ego," that is, in an immanentist way. Morality and religion were its productions and had no transcendental or supernatural derivation.[15]

Feuerbach's influence on Marx can be seen through one key concept: *species*. For Feuerbach, the human species, as opposed to the individual, was generic, ideal, and free from the imperfection or sin that Christianity attributed to it. Christianity, according to Feuerbach, did away with the species, kept only the individual, and substituted God for man. In Christianity, the individual attained his self-centered goal: "Others, the human race, the world, are not necessary to him." Whereas virginity is the ideal of a Christian (Feuerbach upbraids Protestants for abandoning it), sexual love is the ideal of species-being, a love that unites man to the human race (or species).[16]

The third Young Hegelian who had the greatest immediate influence on Marx was Bruno Bauer. A scriptural scholar like Strauss, Bauer obtained a chair at the University of Bonn in 1839. Suppressing the object in the Hegelian subject-object dialectic, he ended up with only the self, or subject. God was only the creation of consciousness, he believed. Bauer accepted the "critical" exegesis that was being carried on in Hegel's wake. Immensely impressed by Strauss's *Life of Jesus,* he soon called himself an atheist. He accepted the proposition that the New Testament was "an alien projection set into a fictional history of Jewish self-consciousness." "By some revolutionary act," Bauer wrote, he headed toward "radical nihilism," leading to "the acceptance of man as God." Hegelianism was losing its liberal Protestant identity and becoming an engine to subvert the "Christian state." Bauer, according to a recent study, was more crucial than Feuerbach to Marx's development up to 1843.[17]

### iii

The unification of Reformed and Lutheran churches in 1817 under the aegis of Frederick William III was a response to the challenge posed to the traditional religious, social, and political order by the French Revolution and by Napoleon. The *Freiheitskrieg* was the beginning of a Prussian effort to establish an agrarian, Protestant, aristocratic, and monarchical regime that was opposed to "the principles of 1789." The fervent, upper-class neo-Pietists strove for the revival of the Protestant church. These Pietists differed from the eighteenth-century ones, because to the former's concern for salvation through Scripture they had added a strong defense of the existing hierarchical order against liberalism and rationalism.

Neither the crown nor the aristocratic Junker landowners had any inhibitions about using political instruments to shore up the church, and vice versa. All church appointments passed through the minister of worship and education and were carefully culled to promote ecclesiastical orthodoxy and political obedience. Hegel was tolerated as affirming these Restoration values, but in the next decade Left, Young, or Neo-Hegelians such as Strauss and Bauer were not. Academic freedom came second to maintaining the Christian Augustinian state.

When August von Kotzebue, a conservative playwright, was assassinated in 1819, Metternich's Carlsbad decrees unleashed a systematic curbing of freedom in the universities. Censorship became routine and the *Burschenschaften* (student societies) were dissolved. The Jewish convert Friedrich Julius Stahl and the philosopher Friedrich von Schelling were appointed to the Berlin faculty to counteract the menace of Left Hegelianism. Ernst Wilhelm Hengstenberg, Protestant pastor who headed the "general staff of Prussian reaction," had married into a Junker family and became the whip of the in Prussia, sounding the hue and cry against heresy. The Hengstenberg Protestant clergy faction, like John Henry Newman's Oxford Movement in England, saw 1830 not only as a political revolution but as a revolution of unbelief as well. The peculiar characteristic of this Christian state was the union of religion and politics—the sacralization of a particular social order and political regime. "Christianity had the responsibility for building God's kingdom in the world, not merely cultivating a form of personal piety . . . [but also a] true church in a Christian state." Unsurprisingly, one policy pursued by the Prussian church-state was opposing the appointment of Jews to public posts unless, like Stahl, they were converts. Indeed, this rigorous policy resulted in over two thousand Jewish conversions in Prussia's old provinces between 1822 and 1840—which did not, however, prevent anti-Semitic incidents. In 1841 a draft of a law proposed reestablishing "corporations" for the Jews in order to protect them and prevent them from "imping[ing] on the Christian state."[18]

This, then, was the backdrop against which Marx's *On the Jewish Question* was written. The Christian state, he felt, was derisory, in that it upheld an inegalitarian social order, which was to be consoled by religion. The Young Hegelians sought the dissolution of this state by denying transcendence and hierarchy and affirming immanence and equality.

Here is how Marx saw it. The Christian state is really not religious but economic. Religious questions can be converted into secular ones, because the aims of religious institutions are secular. For him, there were no transcendental goals; man distinguished himself from other animals by producing his own means of subsistence, not by communicating with the Almighty. Man only *seemed* to have goals that transcended economics.[19]

In several articles in the German press in 1842, Marx attacked the problem of the Christian state. As a dialectician and debunker, it was incumbent upon him to show that the Christian state was riddled with contradictions. Secondly, as an atheist, he obviously recoiled from the whole concept of a city of God and went out on a limb to refute it (something that made him a bit the "theologian" he faulted in Bauer). Detaching ideas from their economic underpinnings was to fall into the trap of the German philosophers whom he would later attack with Engels in *The German Ideology* (1845). In a letter to Arnold Ruge dated March 13, 1843, Marx revealed that his interest in Jewish emancipation was only a means to an end:

> I have just been visited by the chief of the Jewish community here, who has asked me for a petition for the Jews to the Provincial Assembly, and I am willing to do it. However much I dislike the Jewish faith, Bauer's view seems to me too abstract. The thing is to make as many breaches as possible in the Christian state and to smuggle in as much as we can of what is rational.[20]

Marx focused as much on what can be called the Christian question as on the Jewish question. How both these are intertwined with a third problem—the problem of secularism—is of especial interest here.

Marx's polemical technique was to meet his opponent on his own grounds, confounding his confusion. The Christian state, Marx wrote, would unite church and state under a Protestant king. He argued that this union was erroneous on the grounds of St. Augustine's *De civitate Dei,* which he interpreted as having invented the concept of separation of church and state. The Christian state, he suggested, was the offshoot of the concept of the 1815 Holy Alliance, which united Russia, Austria, and Prussia in observing Christian principles, notably to protect the European social order from revolution. Marx curiously cited the pope who, "with profound intelligence and perfect consistence, refused to join on the grounds that the universal Christian link between peoples is the church and not diplomacy, not a secular union of states." Marx embarrassed the Protestant king by citing the pope, whom neither recognized.[21]

Marx's principal point was that the state had no business interfering in religious matters. To do so was fanatical, and flew in the face of all modern political philosophy, which "began to regard the state through human eyes and to deduce its natural laws from reason and experience." He endorsed Enlightenment freedom of secular inquiry, and he attacked the state censorship that had harassed him during his years migrating from city to city (Bonn, Cologne, Paris, and Brussels). In late 1842 he inveighed against the prohibition of divorce and "the polemic against the secular essence of marriage."[22]

Marx drew indiscriminately from a father of the church and modern post-Christian philosophers to promote secularization, or the independence of society from the church. He used the Holy Alliance against the Christian state, neither of which he approved, and invoked the natural rights philosophy; but as a materialist, he could not justify human rights. His overall thrust was to crush the Christian state. The Jewish question could be used to that end as part of the general cause for human emancipation. Marx espoused secularism for both the Jewish people and the Christian state.

To illustrate the different relationships between the state and religion, Marx described Germany, where there was no central state; France, where there was a confused relationship between the state and the Catholic majority; and the United States, where there was a complete separation of church and state—that is, a secularized state where the civil society was unemancipated, being awash in hundreds of religious sects. The United States had achieved political freedom, but not human emancipation.[23]

The Jewish problem attracted the interest of philosophical minds on two fronts: (1) the secularization of the Christian state; and (2) the integration of the Jews into a secular, post-Christian society. In 1843 Bauer published an essay, *Die Judenfrage,* in which he basically threw the onus of emancipation back on the Jews. They were restricted because they had restricted themselves. They had not moved with the progress of history but had demanded that the Christians and the Christian state "give up prejudices," although they remained unwilling to give up their own. The faults with which the Gentiles reproached them were of their own doing. Bauer reviewed the situation of the Jews in Spain and Poland, arguing that much of their misfortune there was their own work. The Jews were too rooted in the past to be accepted into the European present. They must shed some of their culture—their law, for instance, and its casuistry—and adapt to European modes of living. Looking back to Hegelian biblical criticism, he concluded, "Theory has now completed its task, it has dissolved the old contradictions between Judaism and Christianity and can look confidently to History, which pronounces the final judgment on principles which have lost their validity."[24]

Some time later, Bauer authored another essay, "The Capacity of Present-Day Jews and Christians to Become Free." One of its themes was that the onus of liberation should be placed on Christians and Jews alike. The influence of Feuerbach's *Essence of Christianity*, particularly its emphasis on species, was clear. Neither Jew nor Christian, according to Bauer, was true to humanity, in that while one circumcised its children, the other baptized them, and neither accepted their simple humanity. If the Jews wanted to become citizens, they must abandon Judaism. So long as they wanted to remain Jews, Bauer went on, they would never become free. The Christians, for their part, had dissolved "this religious conception of man." Man must now be understood in his "common essence," that is, his universal secular essence. In the post-Enlightenment world, "Christianity dissolves itself, there stands in its place the complete, free man, the creative Mankind hindered no longer from its highest creations." The Jews, Bauer continued, have accepted the Enlightenment dissolution of Christianity better than the Catholics or the Protestants, but they have not applied it to themselves.[25] He explained,

> The most rigid form of the opposition between the Jew and the Christian is the *religious* opposition. How is an opposition resolved? By making it impossible? How is *religious* opposition made possible? By abolishing *religion*.[26]

### iv

In response to Bauer's essay, Marx wrote his response, *On the Jewish Question*, in the fall of 1843 for *Das Deutsche franosische Jahrbucher*, which published it in the spring of 1844. Uninterested in the plight of particular Jews—Rhenish or Polish, for instance—he was concerned with Jews in general, Jews as part of humanity, a nationless, areligious, classless, raceless humanity. His whole approach was specific and particular in his choice of language and images, but abstract in the final analysis. He was not dealing with real interlocutors, the so-called Holy Family (Bruno Bauer, Max Stirner, and Arnold Ruge) or the Utopians (Pierre Proudhon), or Ferdinand Lasalle, who lived luxuriously. The only people Marx did not verbally abuse throughout his long life were his wife, his daughters, and his famous collaborator and creditor, Friedrich Engels. Except for the proletariat, he held humanity—including peasants and Jews—in the utmost contempt. The problem of his time, he thought, was Judaism, not Christianity: "The *social* emancipation of the Jew is the *emancipation of society from Judaism.*" Why?[27]

Marx *did* want Jewish emancipation, but he conceived of that goal in different terms than had previous authors. True to form, he did not conceive the problem as a particularly Jewish one. Marx's Jewish emancipation was *human emancipation* rather than *political emancipation.* The latter derived from the 1789 Declaration of the Rights of Man and the Citizen—the rights of property, speech, religion, association, and due process. These were inadequate because they were concerned only with the rights of the individual. Such rights did not loom large in Marxism (one reason why Marxist regimes could dispense with them in the twentieth century), because they did not address man as a *social* being, a member of a species. To illustrate this point, Marx chose not European countries, whose emancipations were still in progress, but the United States. There, the First Amendment had secularized the state on the federal level. He considered the American state to be "atheist." But he insisted that the "atheism" of the state did not mean atheism in American society, where sects proliferated. One could paraphrase his goal thus: man is not free to practice religion but freed from religion altogether.[28]

The emancipation Marx sought was "an *emancipation of the state* from Judaism, from Christianity, from *religion* in general." Marx's verbal violence exceeded Bauer's: "The state can and must go as far as the *abolition of religion,* the *destruction* of religion" (Marx's emphases). But this destruction of the Christian state could achieve secularization only by "the abolition of private property, . . . the maximum [price controls], . . . confiscation, . . . progressive taxation . . . the abolition of life, the *guillotine.*" Thus, the neo-Pietists, the ministers of religion and culture, and King Frederick William IV were not wrong in seeing the attack on the Christian state as the sign of a more general attack on Christian civilization, or civilization as hitherto known, assuming that life, liberty, and happiness were the general objectives of civilization.[29]

Marx then explained in a truly cryptic fashion that the authentic Christian state did not uphold Christianity. Rather, "the perfect Christian state is the *atheistic* state, the *democratic* state, the state which relegates religion to a place among the other elements of civil society."

This atheistic Christianity, a purely social or humanitarian Christianity, which leaves the supernatural at the bottom of the well, was acceptable to Marx. Arguably, his own humanism had Christian origins. The American separation of church and state left behind its Christian sectarian society for an atheistic government. The state that had not yet done so had "not yet succeeded in expressing the human basis—of which Christianity is the high flown expression—in a *secular, human* form"[30] (all emphases his).

Marx accepted Bauer's and Feuerbach's equation of religion with alienation from the human essence, disagreeing only with Feuerbach's individualistic formulation. He seconded the atheism of both men, but repudiated their nonsocial, individualistic solution. Human emancipation was not what Feuerbach imagined—the return of oneself into oneself—but rather, Marx insisted, the return of man into his *species*. Human society writ large alone could emancipate man completely—religiously, nationally, and ethnically. [31]

> Political democracy is Christian since in it man, not merely one man but every man, ranks as *sovereign,* as the highest being, but it is man in his uncivilized, unsocial form, man in his fortuitous existence, man just as he is, man as he has been corrupted by the whole organization of our society, who has lost himself, been alienated, and handed over to the rule of inhuman conditions and elements—in short, man who is not yet a *real* species being.[32] [emphasis his]

Species-being, that is, is a universal identity, which may ultimately derive from Rousseau's general will, Hegel's universal class (the civil service), or Guizot's middle class (*juste milieu*) that supposedly represented all society. Marx would eventually torpedo these identifying human societies with the proletariat—which, through this logic of numbers, could easily be conflated with species-being. He is opposed to choosing between Jews or Christians, Catholics or Protestants, philosophes or believers because all are less than universal and all beg the question of what it is to be the "essential and true man."[33]

Bauer argued that the Jews could not demand that the Christians give up their prejudices and not give up their own. But Marx thought that men would be equal and free only when they confronted one another as fellow humans, nothing more, and that meant specifically no religion. This is not Christian sociability that, says Marx, revels in the wealth of religious contradictions and diversity. The Marxian universality eschews diversity in favor of uniformity, spelling future disaster for religious, ethnic, ideological, economic, and political identities. This is the meaning of species-being, the foundation block of Marx's future communism.[34]

> Only when the real, individual man re-absorbs in himself the abstract citizen, and as an individual human being has become a *species-being* in his everyday life, in his particular work, and in his particular situation; only when man has recognized and organized his "*forces propres*" as *social* forces and consequently no longer separates social power from himself in the shape of *political* power, only then will emancipation have been accomplished.[35] [emphasis his]

This privileging of the abstract citizen over the real, historical individual is basic training for totalitarianism. The totality has complete sovereignty over the individual—hence Marx's contempt for the ethnic, the sectarian, the factional, the occupational particularity. And here are some other ideological roots of twentieth-century totalitarianism: its bans on labor unions and political parties (other than the Communist Party), and its scorn of ethnic, religious, and political dissidents, in which the Jews played a major role.[36]

Clearly, Marx is shaping a post-Christian universalism to take the place of Christ as the universal shepherd; the lamb who gave his life for all mankind; of St. Paul, who was "all things to all men"; and of the apostles sent to preach to all nations. Marx sees Christianity as divided, sectarian, something that should be replaced by a secular species-being with no religious or ethnic divisions—the nemesis and successor of Christianity.

Bauer broke with Marx because he did not think communism to be the solution to alienation; because the masses would not measure up intellectually to the ideas of the leaders, who would treat them as a collectivity, rather than as individuals. To force feed them simplistic dogmas would defeat the very intellectual liberation for which Bauer and other Young Hegelians had striven. It would lead to a dictatorship, a suppression of freedom, and to inequality, resulting in the few profiteering at the expense of the many—in other words, the very reverse of communism's stated goals. Bauer's critique could not have been more prophetic.[37]

Although Marx was never reconciled to Bauer, he did retain Bauer's identification of religion as an opium and he *did* borrow Bauer's principle of inversion: if man is the subject and God the predicate in ordinary conception, then canceling the predicate and making it part of the subject results in God existing (only) in the thought of man.[38]

But by early 1843, Marx had broken off relations with Bauer and his political solution to the Jewish question. While Bauer's atheism fed Marx's communism, from this point on, Marx drew more on Feuerbach and his concept of species-being.[39]

### V

The title of the second part of Marx's essay was taken from Bauer's "The Capability of Today's Jews and Christians to Become Free." Here Marx vents his spleen, mostly on Jews. The change of tone between the two halves of the essay is symptomatic of Marx's tempestuous and uneven style, and the anti-Semitism is so abusive that it embarrasses his defenders. "The god of the Jews has become secularized and has become the god of the world. The bill of exchange is the real god of the Jew." The Jew is no less sham than the Christian, Marx writes. Jews cannot transcend economic interests in order to acquire the lofty rights of man, which gives them no special claim to emancipation. Does this belief make Marx anti-Semitic, as so many have asserted? Marx seems to view the Jews economically, as he does everyone and everything. He refuses to see them as the "chosen," just as he refused to view Christianity as "spiritual." He secularizes the Jews, seeing in them not the founders of revealed religion, or Christ killers, but capitalists. Marx's Jew is the "everyday Jew," the "huckstering Jew," which is to say, the "secular Jew," the object of Christian anti-Semitism. Rather than ask whether Marx was anti-Semitic, whether he suffered from self-hatred as a Jew, or whether he wrote against Jews not as Jews, but as moneylenders and capitalists, let me suggest one other reading.[40]

Marx wanted to see man stripped of all deceptive identities except those of *homo faber*. The Jew had to become a species-being as much as the Christian. Then he would cease to be a Jew, as much as the Christian would cease to be a Christian. The two would be entirely the same. The common terrain would be simple humanity.

To save the Jew from Marx's putative anti-Semitism, Shlomo Avineri has examined *The Holy Family*, Marx and Engel's first coauthored work, which was published in 1845. There Avineri shows that Marx argued that Jews contributed "to the making of modern times," and that the degree of their emancipation in any given state was a good index of that state's modernity. Marx might have detested the Jews as such, but that did not eliminate their title to emancipation.[41]

What Avineri and others have not mentioned, however, is that Marx in *The Holy Family* is as concerned, if not more concerned, with the issue of secularism as he is with Jewish emancipation. The title, *The Holy Family*, is obviously an antireligious allusion to the Christian Holy Family. Why

"Holy Family?" Probably because Marx had difficulty viewing the Jewish question in completely secular or nontheological terms. Herr Bauer, the atheist, is the "theologian . . . not concerned with politics, but with theology." Bauer even wrote that "the Jewish question is a *religious* question." But Marx does secularize the Jewish question by moving it from the religious to the human sphere. Over-preoccupation with religion meant lack of emancipation from religion—one reason why later generations of Marxists always put religion on the back burner, rather than attack it boldly, as did contemporary fin de siècle anticlericals.[42]

Playing on words, Marx transfers the mystique of the Catholic Sacrifice of the Mass to the Marxian mystique of the "fight of the Mass against the Spirit." Driving home the antithesis between Jewish body-consciousness and Lutheran spiritualism, he writes, "To the material, mass type Jews is preached the *Christian* doctrine of *freedom of the Spirit, freedom in theory* that *spiritualistic* freedom, which *imagines* itself to be free even in chains and whose soul is satisfied with '*the idea*'." For Marx this is Hegelian mystification.[43]

The Jew must come to terms with the Christian state in its "dissolution," not in its prime, Marx states. The Jew can do this only by consenting to his own annihilation as a Jew and his simultaneous self-recognition as a human. The Christian and the Jew meet on the common ground of species-being, one leaving behind the Christian state and the other his "huckstering" existence. Marx prides himself in being party neither to religious delusion nor to money-making (he borrowed instead). His self-image is that of a secular prophet. Since he believed that there is no heaven, there must be justice and happiness on earth. These can be attained only by negating (liquidating?) those who deprive men of it here and now—that is, Junkers and Jews.[44]

The last sentence of Marx's tract is, "The social emancipation of the Jew is the *emancipation of society from Judaism*" (emphasis his). Only the atheistic state unites men as species-beings. The eradication of capitalism, which is the essence of Judaism, liberates society from both. [45]

Marx has committed a capital offense. He has made his own people the prime movers of an economic system to destroy which he will devote his life.

It is hard to accept Marx's tract as representing a "humanism." No groups, not Jews, Christians, kings, theologians, atheists, Hegelians, neo-Hegelians, or utopians, are spared in Marx's omnivorous, dialectical chew. Only the totalitarian society of the species-being is allowed to subsist. It is a foretaste of the even more devastating *Communist Manifesto* (1847) and of its assault on Christian civilization. The emphasis has changed from theology and atheism to the bourgeoisie and the proletariat.

## vi

Hegel's God and man's self-consciousness were too transcendental for Hegel's followers, the "Young Hegelians." Hegel intimated that God had become nothing but a thought in the mind of man, a projection or an alienation. Unsatisfied, each Young Hegelian pushed that logic of secularization a step further. Strauss, Feuerbach, Bauer, and Marx preached a world without God, a world in which God was nothing but a fictive guarantor of a social order, and religion an opium that induced its acceptance. The Kantian assumption of God and the whole issue of material equality paled before other issues, such as moral or spiritual equality or the respect of persons regardless of social standing or origin—the respect of every individual as an end in himself. Not only does Marx not recognize these values, but by not recognizing the spiritual worth of the individual, he unconsciously eliminates all reasons for accepting the communistic material equality. For if the rights of man, including the right to own property, are nonexistent, how can a worker lay claim to the "surplus value" or the property he produces as capital? If he has no property, how can he be expropriated?

The Christian state and Jewish emancipation seem to have been historically incompatible. There was no complete Jewish emancipation before the disintegration of the Christian state, or the old regime, in the French Revolution. The revival of a Prussian Protestant Christian state in the middle decades of the nineteenth century paired the Jewish question with the menace of rationalism and revolution emanating from France.

Marx abhorred the Christian state, but he did not for that reason champion Jewish emancipation. Rather, he wished to substitute for both the concept of species-being, or secularism—the common denominator of members of the species. Private property and religion, on the other hand, serve to particularize men, as do ethnicity and nationality. Embracing the concept of species-being, seeing it as the bedrock of human solidarity, naturally spells communism, because communism strips us of all that is proper to us, as individuals, as members of social classes, or as estates, and unites us to all that we hold in common. Secularism is an attribute of species, and its opposite is religion. Secularism would replace Christian universalism, which had lost ground to Reformation secularism and Enlightenment rationalism.

The emergence of the Jewish question, the driving force of this last stage of secularization, was a sign and a symptom. The real campaign for secularization began with the Enlightenment, which was not uniformly Semitic. Nor was Marx. The battle for Jewish emancipation was adopted inter alia by many of the leading minds of the late eighteenth and nineteenth centuries—possibly more in the former than in the latter. In the case of the Young Hegelians—of whom Marx was one, despite his vituperations against them—Jewish emancipation was only a corollary to their atheism and their campaign to abolish religion. It was one of several issues of the day, which included the secularization of the state—education, marriage, vital statistics, the Sabbath—and the abolition of religious tests, emblems, symbols, and ceremonies.

One of these issues concerned the Jews in particular—the abolition of religious tests and equal access to public office. There were undoubtedly several reasons for these tests, but one would be the fear that non-Christians would have a deleterious influence, through the spread of anti-Christian ideas, on the minds and souls of a public and government that was still Christian. If toleration meant equal access, then a race of people with superior intelligence and motivation could eventually marginalize Christianity in the public sphere. Marx's commitment to the equality of species, on the other hand, would hinder any kind of ethnic elitism, or even survival. Hence one source of anti-Semitism was Marxism.

Both Marxism and Nazi racialism have been far less benign for the contemporary Jew than the antecedent Christian civilization, however deplorable the latter was at times. The Western liberal, secular state of the twentieth century may be an improvement in several respects over the Christian state, but the cost it exacted—the seclusion and exclusion of racial and religious identities—is harsh for Jews as well as Gentiles. The survival of the Jewish religion and Jewish people is threatened by the culture of the non-Jewish Jew, which is but another variety of secularism.[46]

# Part V
## Introduction:
## Reaction against Secularism

The destiny of everything Western was to penetrate and dominate almost every other corner of the globe in the second half of the nineteenth century. Imperialism brought colonial administration, extracted raw materials, invested capital, and supervised industrialization. These vast movements are well known. Less understood is the impact of modern European thought beyond its borders, which may have entailed even profounder effects.

Our sharp eastward turn from Prussia (Kant, Hegel, Marx) to Russia is of course justifiable by the sharp eastward turn of revolution and Marxism. There is no question in the minds of the two men studied in this concluding section that the apostasy and nihilism of the West spelled disaster in the East. But both men offered hope by diagnosing the malady.

For decades, Russia had debated the relative benefits of westernization versus Slavophilism. Part of this debate was inspired by nationalism: Why did Russia need to look beyond its borders for enlightenment, particularly when what was beyond looked so formidable and foreign to Russia's spiritual traditions—materialism, Hegelianism, scientism, evolutionism—in other words, nineteenth-century secularism? Dostoyevsky saw in these recent trends the ingredients of nihilism and the seeds of something indefinable, which Aleksandr Solzhenitsyn discovered later to be totalitarianism. Solzhenitsyn was born in 1918, a little more than a generation after Dostoyevsky's death in 1881. These two thinkers reacted against the extinction of traditional values.

A left-wing sympathizer of revolution in 1848, Dostoyevsky swung around to Christianity after imprisonment in Siberia. He then traced the ultimate, ineluctable consequences of moral nihilism in three novels, most notably in *Demons* (1871), which we analyze below. Solzhenitsyn's *Gulag*

*Archipelago* recounts his eight years in Stalin's Siberian labor camps, where millions died. In 1975 he was exiled for having published this work abroad and thus disgracing the Soviet Union. At Harvard in 1978, he shocked and angered many Americans by suggesting that something similar to the Gulag could happen to us, given the deep flaws in our civilization.

Americans, however, usually accept criticism generously ever since Alexis de Tocqueville published his *Democracy in America* (vol. 1, 1835; vol. 2, 1840), of which there are some echoes in Solzhenitsyn's speech. Hopefully we can once more prove to be resilient—spiritually as well as materially (in which we so excel).

# Chapter Twelve
# Dostoyevsky and European Secularism

Fyodor Dostoyevsky (1821–1881) returned to St. Petersburg, after nine years of imprisonment and exile in Siberia for his involvement in the revolutionary Petrashevsky circle, convinced that Russia's true identity lay with the peasants he had grown to understand and love in exile, rather than with the westernizing Russian intellectuals. Russia, he thought, should realize its own greatness rather than follow Europe's. This meant embracing Russian Orthodoxy and autocracy rather than Roman Catholicism and European socialism, both of which he would reject vehemently in *The Brothers Karamazov*.[1]

One of the great nineteenth-century novelists, Dostoyevsky was brought up on a rural estate and trained as a military engineer, but in 1859 he forsook a comfortable career in the civil service for a penurious one as a serial novelist in various literary journals. Though dealing "with real bodies and blood," Dostoyevsky hardly shunned ideas. Indeed, *Demons* could be called a political novel with metaphysical ambitions, for it deals with the deleterious effects of Enlightenment and Hegelian secularism on serious Russians, who took it to its logical conclusions. In it, the formula of *The Brothers Karamazov*—if there is no immortality, everything is permitted—is evident. Ideas are materialized by convincing, and often extreme, characters. These are swayed by the latest waves of European thought, following in the footsteps of Ivan Turgenev's *Fathers and Sons* (1862) and Nicholas Chernyshevsky's nihilist novel, *What Is to Be Done?* (1863). But rather than being an exponent of secularism, Dostoyevsky is one of its major critics, although he never hesitates to criticize a major religious denomination, too.[2]

Dostoyevsky's literary strategy demonstrates nihilism in extremis. As sadism has been said to be the logical consequence of Enlightenment hedonism in Marquis de Sade's France, so is nihilism, for Dostoyevsky, the terminus of nineteenth-century revolution. Nihilism can erupt as murder, or as an arbitrary marriage to a "crazy cripple" (Nikolai Stavrogin to Marya Lebyadkina in *Demons*) to spite a mother who had raised him in the highest imperial fashion. In Nikolai, Dostoyevsky creates someone who acts without rational motivation, defying the logic of the Enlightenment and the mores of polite company. Nikolai is a man contemptuous of the ways of the upper classes, a revolutionary who does not hesitate to act crudely and cruelly in order to *épater les bourgeois*.[3]

Nihilism can take place within the family and the town, and amid normal social relations, which are systematically destroyed. It is a successor to and consequence of the Enlightenment.

i

The antidote to nihilism, Dostoyevsky believes, is Slavophilism—a rejection of the utopian and nihilistic revolutionism of the West. "In many respects," Dostoyevsky writes in the summer of 1877, "I hold Slavophile convictions, even though I am not quite a Slavophile." To oppose westernization, he evokes the "great Eastern eagle, shining with his two wings on the peaks of Christianity, soar[ing] over the world." Two experiences shaped him and brought him to this conviction: his rural childhood among the peasantry on his father's lands and his Siberian exile among peasant convicts.[4]

As a young boy, Dostoyevsky used to play in his father's forests in Darove. One day, at age eight or nine, he thought he heard a wolf nearby and ran terrified out of the woods into the arms of a peasant named Marey, who embraced him, made the sign of the cross on his face, and restored his calm. This was something of a revelation for Dostoyevsky—a revelation of the peasants' deep-rooted faith and fundamental kindness.[5]

Dostoyevsky's second insight into the peasant character occurred during his Siberian exile during the 1850s. He was initially turned off by the peasants' vulgarity and crudeness, both physical and moral; they were constantly embroiled in drunken brawls. But underneath this brutish behavior lay the sincerity and genuine belief of the peasant soul. The peasants, he discovered, were not in the least interested in social equality; they preferred to keep Dostoyevsky at a gentlemanly distance, just as Dostoyevsky kept himself from them. But they did have a culture, one based on faith. Dostoyevsky's conversion came one day when, after witnessing drunken peasants carousing, he went back to the barracks and thought of the peasant Marey. Even earlier, he ruminated, his nurse had inducted him into the Christian martyrology of peasant culture. His contacts with the peasants of

Siberia brought about a transformation to an "entire psychic-emotive equilibrium." *Pochvennichestvo,* or return to the soil, became Dostoyevsky's solution to Russia's problems. What he had previously sought in Western culture, he now found "embodied in the instinctive moral reflexes of the much-despised and denigrated Russian peasant."[6]

After the Crimean War, a war waged against Russia by France and Britain in the 1850s, many Russian intellectuals turned patriotic. But the sea change went beyond that. As Joseph Frank, Dostoyevsky's extraordinary biographer, observed, Russia, "having steeped itself in European culture, realizes that it has lost its native roots and accordingly turns back on itself with destructive skepticism." With *Demons,* Dostoyevsky hoped to force people to recognize a purer ideal: "Christianity is the Russian Land's only refuge from all its evils. I pray to God that I manage [to convey] it."

The West had inspired the Russian elite for over a century and a half, but it could no longer play that role, Dostoyevsky felt, because it had lost its Christian identity. The Roman Catholic Church "has long ago sold Christ for earthly rule . . . [and] has naturally generated socialism." The Grand Inquisitor in *The Brothers Karamazov* (1879/80) ruled over a flock of submissive believers. Their church had succumbed to the devil's Third Temptation, which Christ rejected: choosing earthly rule over the kingdom of God. Socialism as a secularization of Catholicism, Dostoyevsky believed, had stepped into the shoes of St. Peter, offering bread in exchange for freedom and the kingdom of heaven. Dostoyevsky came to hate both Catholicism and socialism, and for the same reason: both, he felt, enslaved man by subjecting the supernatural message of Christ to earthly power—in other words, to secularism.[7]

> In the West Christ has been lost [through the fault of Catholicism], and because of that the West is declining, exclusively because of that. The ideal has changed, and how clear that is! And the downfall of papal power along with the downfall of the head of the Romano-German world [France and its friend], what a coincidence![8]

Dostoyevsky was no better disposed to Western European Jewry than to Roman Catholicism. In Russia, he argued, Jewish moneylending was responsible for the usurious subjugation of the peasantry. "Germany is becoming terribly Judaized in all respects," he asserted; and Jewish influence was linked to the twilight of Christianity in the West. "We are speaking about *Judaism* and the *Jewish idea,* which is clasping the whole world instead of Christianity, which did not succeed." Intellectuals, he thought, had come to side with the "Jews," because of their common hatred for Christianity.[9]

Dostoyevsky, however, waxed enthusiastic about Bismarck's Germany, because of its anti-Catholic and antisocialist policies. But a further step was needed: Germany needed to free itself from the Western yoke and join Russia. Together they could save the world.[10]

A counterweight to this pessimism about a declining Christian Europe—a pessimism that was reinforced by a miserable (self-imposed) exile in Europe to evade his Russian creditors between 1867 and 1871—was the view that Russia and Russian Orthodoxy were all Europe's best hope. But so negative was Dostoyevsky's view of "Europe" that one scholar argued that Dostoyevsky's four years in "Europe" were "a purgatory of the gaming table and debt and epilepsy and [an] enforced exile among people who were in many ways more distasteful to him than the convicts in Siberia." Russia and Russian Orthodoxy alone could save Europe from the Babylon it had become, for they alone shrugged off public opinion.[11]

After this "sudden invasion of unbelief," Russia was faced with the option of accepting Western apostasy (from the left-wing Hegelians Feuerbach, Stirner, Strauss, Bauer, Marx, and others) or pushing it to its logical extreme. Dostoyevsky may or may not have read Hegel, but a Hegelian current runs through Alexander Herzen and the well-known literary critic Vissarion Belinsky, whom Dostoyevsky had once admired. Indeed, Stepan Verkhovensky, one of the leading characters in *Demons,* is modeled on a Hegelian historian. Stepan is "a repository of the beliefs of the men of the forties in Russia—the liberal intellectual Westernizers."[12]

The Christian Dostoyevsky of the 1860s and 1870s believed that Orthodox Russia should become the third Rome and convert the entire world. The foundation for this worldview came from an irrational personal credo: "If someone proved to me that Christ is outside the truth, and that *in reality* the truth were outside of Christ, then I should prefer to remain with Christ rather than with the truth." Dostoyevsky made this extreme criticism of secularism in his popular *Diary of a Writer,* which was serialized in the 1870s, and in his black novel, *Demons,* to which we now turn.[13]

## ii

One of the three main characters of *Demons,* Nikolai Stavrogin, in his confession to the Orthodox monk Tikhom, expresses the demoralization that is stereotypical of the westernizing intellectual: "And not only have I lost the sense of good and evil, but good and evil really do not exist." Plumbing the depths of an atheist's soul, a man who discovers nothing, is the subject of the novel's most devastating chapters. How did Dostoyevsky get there?[14]

The reverberations of the Parisian revolution of 1848 triggered imitations in most European capitals. It seemed like "a springtime of hope" to Russian intellectuals. "The old order of things is shaking and breaking up," Dostoyevsky wrote. Petrashevsky's westernizing, revolutionary circle, to which Dostoyevsky belonged from 1846 to 1849, was inspired by the French utopian socialist Charles Fourier, who sought to replace private property with "phalansteries," or communes, grouping together 1,620 persons of compatible temperaments. Nicholas Speshnev, a communist member of the circle, envisioned the phalanstery as a possible socialist model to replace the traditional Russian peasant commune. Moreover, Petrashevsky's secret society was headed by a central committee, a model for the one in *Demons*.[15]

Nihilism is the background of Dostoyevsky's *Demons*. Before this revolutionary movement reached its apex in the late 1860s, Ivan Turgenev, a leading noble and a westernizing author, who was earlier influenced by Michael Bakunin, published in 1862 a shocking portrait of a nihilist in *Fathers and Sons,* or *Fathers and Children,* as it is sometimes translated. Turgenev's protagnist, Bazarov, a doctoral student in the sciences, believed that "all people are alike and it is not worth while to study them." He ridicules or patronizes the elder gentry and their servants. Presumptuously, he pits the old classical learning against the new, superior laboratory knowledge. He also denounces "feudal parsonages." Arguing that "love . . . is a purely imaginary feeling," he falls in love with an aristocrat. Like Dr. Homais, the pharmacist-exponent of modern medicine and the Enlightenment in Flaubert's *Madame Bovary* (1857), who ended up treating a patient's foot so badly with his advanced knowledge that it had to be amputated, Bazarov himself dies of typhus after dissecting a victim of the disease. Both novels undermine the pretentions of modern science as the answer to the human condition.[16]

Dostoyevsky wrote to Turgenev praising his novel, which contrasted very effectively the older generation of Orthodox Christians with the "new men"—atheists who tore apart sacred conventions. Indeed, Dostoyevsky may have gotten his inspiration to write a "book about atheism" from this little antinihilist novel, which concluded with the thought that the prayer of the parents for their deceased son was not effective.[17]

A second book that had great influence on Dostoyevsky is Nicholas Chernyshevsky's "novel" *What Is to Be Done? Tales of the New People* (1863), which was written while the author was in Peter and Paul Fortress, a year before his final exile to Siberia. Exceeding Bazarov, Chernyshevsky's characters are too reduced, too simplified. They attempt to understand aesthetics in scientific terms and believe that "art should merely reproduce reality" for the sake of utility. Morality, he felt, had to be altruistic to be truly moral.[18]

The novelty of *What Is to Be Done?* is its espousal of the emancipation of women and workers and the reform of marriage. It all takes place in a textile cooperative, where the profit motive and private property have been largely banished. The subordination of personal happiness to a social cause—the sacrifice of a husband's career and the couple's libido (they never have intercourse), and finally the wife's infidelity—leads the husband to self-destruction. Humanity is annihilated when Vera, the wife, remarries a week after her husband commits suicide.

Dostoyevsky was not opposed to emancipation in either the domestic or the social spheres, but he believed that freedom could not be had in purely materialistic terms. It could only come about as a result of free will. Scientism, determinism, and socialism were all abhorrent to his libertarianism. In *Demons,* the character Lebyadkin seems to express Dostoyevsky's view about Vera's kind of scientism: "The whole spirit of these last few centuries of ours, taken as a whole with its scientific and practical emphasis, is perhaps indeed damned, sir!"[19]

The immediate impulse to write this novel about nihilists was the murder of a student, named Ivan Ivanov, in the Moscow School of Agriculture, on November 21, 1869, probably by Nechaev, the head of an anarchist group (Narodnaya Rasprava), or by one of its members. This cold-blooded murder, inspired by Nechaev's nihilistic credo, horrified the Russian public, which feared that killing might become epidemic. Dozens of suspects were arrested. Nechaev escaped to Switzerland, leaving behind his *Revolutionary Catechism* (possibly the work of Bakunin), which heightened the general alarm. Nihilist revolutionary cells—such as Chernyshevsky's "Land and Liberty" with its "four man plus one" leadership, symbolized by a fist of four fingers plus one thumb—were probably the models for Dostoyevsky's nihilist cell, the Quintet, in *Demons.* The following are some of the apodictical utterances of Nechaev's (or Bukharin) *Revolutionary Catechism*—a classic in revolutionary absolutism:

1. The revolutionary is a lost man; he has no interests of his own, no cause of his own, no feelings, no habits, no belongings; he does not even have a name. Everything in him is absorbed by a single, exclusive interest, a single thought, a single passion—the revolution.

2. In the very depths of his being, not just in words but in deed, he has broken every tie with the civil order, with the educated world and all laws,

conventions and generally accepted conditions, and with the ethics of this world. He will be an implacable enemy of this world, and if he continues to live in it, that will only be so as to destroy it more effectively.

3. The revolutionary despises all doctrinairism. He has rejected the science of the world, leaving it to the next generation; he knows only one science, that of destruction.

4. He despises public opinion; he despises and hates the existing social ethic in all its demands and expressions; for him, everything that allows the triumph of the revolution is moral, and everything that stands in its way is immoral.[20]

This nihilism, whose exclusive mission is to destroy, negates science as well as religion. Christianity saves; nihilism, true to its name, can only destroy.

Dostoyevsky was shocked by Nechaev's murder of Ivanov, which he learned about in Dresden in December 1869. He began keeping material for his *Notebooks,* in a chapter titled "The Devils," in which he would record bits of conversation of nihilists and antinihilists for the novel he had thought to write about atheism. In his mind, criminality and unbelief were linked, as indeed they seemed to be in the *Revolutionary Catechism.* But deeds also depended on what kind of belief a person held: Roman Catholicism, thought Dostoyevsky, was anti-Christ; Judaism was usury; and atheism was nihilism, for it believed in neither good nor evil, as Nietzsche explained a decade later. The latter contradicted the view of the seventeenth-century Protestant philosopher Pierre Bayle, who claimed that people did not act, necessarily, in consequence of their belief. Atheists could therefore be virtuous, Bayle argued, and the pious could commit crimes—a view that the Enlightenment welcomed. Dostoyevsky, for his part, maintained that nihilism, and therefore communism, could be accounted for not just by external circumstances, but by individual choice. *Demons* was a prophetic work, as it has often been described. If secularism is not just restricting religion to the private sphere but superseding it by unbelief, as proposed by Nechaev and illustrated by two of Dostoyevsky's characters in *Demons,* Pyotr Verkhovensky and Nikolai Stavrogin, then Dostoyevsky is definitely writing about secularism, but a more radical secularism than hitherto seen. In the *Notebooks,* begun in early 1870, he gives a glimpse of the novel that appeared a year later:

> Nobody in Russia knows who he is. We have lost sight of Russia. We can't recognize our own peculiar nature, nor do we know how to deal independently

with the West . . . this is a matter of the ultimate results of Peter the Great's reforms. He decided that his Russians must become Europeans by decree, and 150 years later he's finally got them, his Europeans. To be sure, they've become alienated from their own people and yet haven't become attached to any other nation—because the others are all national, whereas we deny nationality on principle, wanting to be just Europeans, although there just isn't such a thing as a European.[21]

For Dostoyevsky, Russian Orthodoxy is the sacred flame of Russia; Peter the Great was the first westernizer and nihilist who separated the people from the intelligentsia. All that remained of Russia was the holy remnant, the peasantry and the Orthodox Church, which preserved the purity of Christian belief. He went on,

> We shall smash those European fetters . . . and we shall all realize, finally, that the world, the terrestrial globe, the Earth has never *seen such* a gigantic idea as the one which is now taking shape here, in the East, and moving to take up the place of the European masses . . . to regenerate the world.[22]

The prince, Nikolai Stavrogin, in the early pages of the *Notebook* for *Demons*, is Orthodox, a man who prays before icons and believes that "the Russian idea will save mankind." He should be painted, Dostoyevsky muses, as "an enemy of nihilism and liberalism, and as a haughty aristocrat." For the prince, belief in God is belief in Russia. "Only Russia can preach Christ. Only the Russian people are a God-bearing people." Yet somewhat unpredictably (but characteristically in the novel), he sides with Nechaev, agreeing that one must "burn everything." Fire is the symbol of revolution, destruction, and nihilism. Even burning babies, he thinks, is permitted for the sake of population control. Fire was the weapon of the workers' insurrection in the Paris Commune of 1871. Paris was burning even as Dostoyevsky wrote. Bakhtin, the Russian critic of the 1920s, believed that the Dionysiac festivals of fire were the spontaneous creations of the people. Lenin believed that festivals of the masses were actually revolutions.[23]

Nechaev, the nihilist, averred that the "Christian religion must be totally eradicated, so that one could start a new life . . . without any kind of God." Pyotr, a character in *Demons* who models himself on Nechaev, states that this was "enough reason to shoot oneself." Pyotr endows the Antichrist of the Book of Revelation and others with a principal trait: Arianism, or the denial of Christ's divinity. According to Geir Kjetsaa, Arianism was the precursor of secularization. In Dostoyevsky's words, "The Antichrist has already been born . . . and he is on the way."[24]

In the *Notebooks,* Dostoyevsky argues on two planes. First, on the national level, Russia can be seen as the only remaining torchbearer of Christianity and with it she "must free herself from the German and Westernizing yoke, and become herself." Second, at the individual and moral level, expressed by the Russian Orthodox Bishop Tikhon, "If there is no God, how will your world, and you in it, stand up, even for a single minute? Why do good deeds, if this is so? Why sacrifice oneself?" In short, secularism is the abyss of Western materialism, an apostasy marking both the individual and the nation. Dostoyevsky plumbs deeper, saying that "the main idea of socialism is *mechanism* . . . socialists are wild about the notion that man himself is nothing more than mechanics."[25]

*Demons,* or *The Possessed,* was published serially in the *Russian Messenger* in 1871–72. The setting is in an unspecified provincial capital near "Skvoreshniki," the country estate of one of the principal characters, Varvara Petrovna. *The Possessed* anatomizes this provincial town from top to bottom, from gentility to the newly freed serfs, emancipated by Alexander II in 1861.

The question Dostoyevsky seems to pose is, What happens to a small provincial capital that has been infected by the poisons of Western modernity?

Stepan Trofimovich Verkhovensky is arguably *the* central character of this long novel (although Dostoyevsky denied it), which was designed to be read, perhaps, in the long winter nights by the upper ranks of Russian society. Indeed, Stepan was educated in the West and constantly parenthesizes his conversation in French—the language of the westernizing aristocracy. He was the tutor of Nikolai, Varvara's son, now a grown man, and lives under the patronage of this wealthy general's widow and contractor's daughter at his country estate. Having spent time in Switzerland, which harbored revolutionary exiles from 1848 to 1917, Stepan lives on imaginary laurels and is constantly afraid that he has been forgotten in high literary circles, where he was probably never remembered. Dostoyevsky, in painting this weak, foppish, but likeable character—in contrast to the nihilist murderers and suicides in the novel—continually points out the irony between his claims to fame and his actual insignificance. Commenting on Stepan's illusion of having been on a level with Herzen many years ago, the author writes, "But Stepan Trofimovich's activity ended almost the moment it began." Stepan represents a *philosophe* without will, a man who lives entirely in the realism of scattered ideas. His eighteenth-century character enables him to love knowledge for the sake of knowledge, which he stretches to absurd lengths of pedantry and antiquarianism. While he lives in the past, he poses as very avant-garde because he is interested in all the intellectual novelties emanating from Western Europe. There is "something very indefinite" about a poem he wrote. His interests range from science, about which Dostoyevsky cannot resist another jab—he is "a scholar, however . . . well, in a word . . . he did very little as a scholar, nothing at all"—to history; he harped on such useless trivia as the unfulfilled greatness of a Hanseatic town in the Baltic between the years 1413 and 1428. He was brought low, as he saw it, by "circumstances" rather than by any personal weakness.[26]

Stepan does not practice any religion, yet he is not an atheist. It is believed that Dostoyevsky used the Russian Hegelian historian Timofei Granovsky, a nonrevolutionary westernizer close to Herzen, as Stepan's model. Like Victor Hugo, Auguste Comte, and the neo-Hegelians whom he imbibes, Stepan has found religion in the cause of humanity—that is, in everything "progressive" of the century. His theism is Hegelian: "I believe in Him as a being who is conscious of himself in me," he explains. Like Granovsky, he has read Feuerbach's *Essence of Christianity*, which places him amid believers in the general dissolution of Christianity. "For all my sincere respect for it," he explains, "I am not a Christian. I am rather an ancient pagan, like the great Goethe, or like an ancient Greek." If he is a pagan, he must be a great one! Such an orientation is far more respectable in the circles he once frequented—those of the 1840–1848 generation—than in some naive orthodoxy. Perhaps he could be described as Dostoyevsky described his Underground Man: "consciousness is inertia."[27]

Such a weak man is naturally under the control of the novel's main female character, Varvara Petrovna. It is at her estate, once the home of several hundred serfs about whom we hear little, that much of the novel's action takes place. When a baron friend of Varvara mentions the imminent emancipation of these serfs, Stepan shouts an inappropriate "Hurrah!" which merits her steely glower. In fact, Varvara has Stepan wholly under her thumb. Stepan has no autonomy, and he depends upon Varvara to nourish his imposture. Indulgently, she refers in public to his imaginary fame. His relationship to her is quite like that of a minor philosophe to an eighteenth-century *salonnière*. Dostoevsky elaborates:

> She protected him from every speck of dust, fussed over him for twenty-two years, would lie awake whole nights from worry if his reputation as a poet, scholar, or civic figure were in question. She invented him, and she was the beehive in her invention . . . She herself even invented a costume for him, in which he went about all his life.[28]

In keeping with this relationship of mistress-philosophe, Varvara decides that they must go to the capital: "It was decided to go to Petersburg without a moment's delay, to find out everything on the spot, to go into everything personally, and if possible, to throw themselves heart and soul into the new movement." They will even found a magazine.

Their endeavor is thrown off by the revolting vanity and boorishness of the St. Petersburg intellectuals. When Varvara benevolently offers them her largesse, "she was accused to her face of being a capitalist and an exploiter of labor." She and Stepan are reactionaries, so the avant-garde desert them. Stepan then resolves to go to Berlin, where he believes he will find more receptive minds. He writes from Berlin:

> In the evenings I converse with the young people till dawn, and we have almost Athenian nights, though only in terms of refinement and elegance; it is all quite noble: there is a lot of music, Spanish airs, dreams of universal renewal, the idea of eternal beauty, the Sistine Madonna, a light shot through with darkness, but then there are shots even on the sun![29]

One of Dostoyevsky's strengths as a thinker is his directness. Hegel needs to be read in his entirety to ascertain whether he is a theist or an atheist, to understand him and still misunderstand him. But Dostoyevsky, in the midst of writing *Demons,* reveals that "the main question, which is pursued in all the parts, is the same one that I have been tormented by consciously and unconsciously my whole life—the existence of God." *Demons* is an attempt to illustrate what happens to a society when its intelligentsia loses belief in God. Dostoyevsky confesses that he himself had been drawn to atheism and socialism, and eventually to nihilism, in the 1840s by a Westerner, Vissarion Belinsky. But shedding such youthful aberrations, Dostoevsky shows how the self-interest or utilitarian morality of Jeremy Bentham leads to reprehensible consequences. He does this in his character Karmazinov, the famous writer in the novel, who is a caricature of Turgenev. Dostoyevsky goes on to catalog the negative contributions of the nineteenth-century West in the father of positivism, Auguste Comte, and the neo-Hegelian Max Stirner. He also takes on a Romantic utopian, Charles Fourier (whose "tarnished eighteenth-century faith . . . [is] so unsuited for our soil"), the atheist Feuerbach (whose *Essence of Christianity* devastated "our Western circles . . . and rapidly obliterate[d] the remnants of all preceding outlooks"), and Victor Hugo (whose "work . . . [is a socialist] justification of the pariahs of mankind [that] provided the moral foundations of the modern world"). One way or the other, they have all "rejected God," leading to the worship "of humanity," the common denominator for all these thinkers. Everything points to an actual deification of mankind.

By 1870 the state of European culture, Dostoyevsky writes, was such that "there is no truth, there are no truth-seekers; atheism, Darwinism, Moscow church bells." Dostoyevsky believes that it is his mission to take to extremes the floating premises of the European intelligentsia. "No true union is possible between the Christian Russian people and the embodied essence of godless Russian Europeanism." Stepan, then, is the not-so-innocent character who has infected Russia with European devils.[30]

Stepan's infatuation with these latest ideas is tinged with apprehensions about their effect on the next generation. "He fears the provincial governor takes him as a corrupter of youth and a fomenter of provincial atheism." On reading *What Is to Be Done?* (1863), he begins to take stock of where things are going:

> We were the first to plant it, to nurture it, to prepare it, and what new could they say on their own after us? But, God, how it's all perverted, distorted, mutilated! . . . Are these the conclusions we strove for? Who can understand the initial thought here?[31]

Stepan is the problem Dostoyevsky wants to anatomize. For starters, he is dominated by Varvara, just as the governor Lembke is dominated by his wife, Yulia Mikhaylovna, and as the revolutionary martyr Shatov is dominated, briefly, by his ex-wife, Marie. Even in 1870, Dostoyevsky was attuned to revolutionary feminism.

### iii

*Demons* contains three parts. Parts One and Two focus primarily on Stepan and Varvara and their provincial world. Part Three centers around a festival, or fete, organized by Varvara's rival, Yulia Mikhaylovna, who had had a fortune in serfs and is the wife of the newly appointed provincial governor, Andrey Antonovitch von Lembke. This part climaxes with the murder of Shatov, the ex-serf, by Stepan's son, Pyotr, and with the suicide of Varvara's son, Stavrogin. The fete dramatically separates the two parts of the action: the older generation's aspirations for reconciliation, recognition, and enlightenment against the younger generation's conspiracy, subversion, and self-destruction.

Undaunted by the fiasco of her and Stepan's literary jaunt in St. Petersburg, the strong-willed Varvara decides to cosponsor Yulia Mikhaylovna's country festival, an outdoor literary fete featuring readings by Stepan and by the rather famous novelist in residence, Karmazinov, who is depicted as vain but not silly. The whole town would attend.

The buildup for the festival-fete is considerable. Dostoyevsky throws in all sorts of incongruities: indecent photos dropped into a missionary woman's bag by an "idle" divinity student, a raid of Stepan's quarters by the police, and more. Dostoyevsky delights to point out: "Every little bone in him is aching with delight now; he's never dreamed of such a gala performance," for nothing better authenticated his imaginary revolutionary credentials.[32]

The first presentation at the festival is an insolent poem read by one of the revolutionary Quintet, Liputin, on the marriage prospects of the governesses, who were the beneficiaries of the fete. Karmazinov's reading is attended by throngs of people, "the rabble" of the town, who fail to understand his poetic license and insist that the ghosts he alludes to do not exist and that the world does not rest on three fishes. Karmazinov takes offense that his symbolism escapes them completely. Stepan's lecture is even more baffling: "And I proclaim that Shakespeare and Raphael are higher than the emancipation of the serfs, higher than nationality, higher than socialism, higher than the younger generation, higher than chemistry, higher than almost all mankind." Openly embarrassed by a divinity student, he bursts into sobs. The last speaker, tapping into national pride with an accusation about the international disgrace Russia has fallen into, is applauded wildly. But the festival is a literary and social fiasco symbolized by the fire that consumes it.[33]

First thought to have been set by workers from the factory, the fire is subsequently blamed on the ex-convict and ex-serf Fedka. Ironically, it destroys some of the workers' dwellings down by the river—that is, those on the bottom of society's rungs do not benefit from nihilism. While the fire is burning "in people's minds"—a metaphor for the incendiary fete and its revolutionary potential—Captain Lebyadkin and Marya, Stavrogin's wife and the captain's sister, are murdered in their lodgings apparently by the same convict, Fedka, and another ex-serf who set the fire with two workmen. The motive is apparently common theft. Crime and revolution are thus intertwined.[34]

The purpose of the fete was to unite all classes by bringing high culture—both literary and political—to the level of the people. This aim fails dismally: half the trades people are absent, whereas the other half jeer at the readings. An inebriated crowd interrupts the ball of the second half of the festival. Everyone anticipates trouble. Yulia Lembke becomes more and more embarrassed, and her husband, the governor, becomes less and less able to contain the disorder. Musicians are beaten, and shocking dances are performed. "By morning," Dostoyevsky writes, "they were drinking to distraction . . . and only at dawn did part of this rabble . . . arrive at the scene of the dying down fire for new disorders." One critic has pointed to the fete as a kind of medieval bestiary of deformed characters—"the fete of fools and of carnivals," as he put it. Humanity, left to society, resembles the grotesque more than the sublime. Indeed, the learned—Karmazinov, Stepan, and Lembke—did make perfectly grotesque fools of themselves. The pandemonium generated by the fete, Dostoyevsky seemed to say, mirrors the Nechaev affair with its "student uprisings, literary scandals and the St. Petersburg fires."[35]

Fire ignites the minds of men with revolution. Dostoyevsky calls it "a new religion . . . on its way to replace the old one," and says that "this is a big thing." It destroys everything in its path. In the book, the governor himself is hit by a smoldering beam while trying to assist the firefighting. His wife is devastated by the drunken brawl into which her festival has degenerated.[36]

The stunning scene allowed Dostoyevsky to rise to the peak of his creative powers; it might have issued ultimately from the epileptic fits he was currently suffering. In the excitement of this double fire (intellectual and physical) he delves into the nihilism of the revolutionary cell, the Quintet. This organization is not known to the governor; even though he has collected revolutionary manifestoes from all parts of Russia, this one mushroomed under his nose. He is unable either to interpret it correctly or to recognize its real-life versions. Karmazinov, Dostoyevsky tells us, is better informed and considers Pyotr, " if not the ringleader of everything covertly revolutionary in the whole of Russia, at least one of those most deeply initiated into the secrets of the Russian revolution." Dostoyevsky's use of the "Russian revolution" illustrates his *prévoyance*. Karmazinov, while explaining why he is leaving Russia (like a rat leaving a sinking ship) predicts that "Russia now is pre-eminently the place in the whole world where anything you like can happen without the least resistance." What distinguishes the Russian revolutionaries from their Western counterparts is "the unheard of boldness in looking truth straight in the face"—something that "belongs only to the Russians of this generation." Its strength was putting into effect what were otherwise merely liberal conversation pieces.[37]

But revolutionary will is in fact the will of only a tiny minority. When Pyotr says, "We are going to make a revolution," he is talking about five people, not the national network of cells he brags about. Dostoyevsky, however, observes that this "vanguard" can do a lot. The polite society of the towns did not heed warnings about a revolution. "While the government purposely gets the people drunk on vodka so as . . . to keep them from rebelling," the authority, the governor, throws up his hands exclaiming, "It's all arson! It's nihilism! If anything's ablaze, it's nihilism!"[38]

The fete is the moment when the Romantic generation of the 1840s goes under and the nihilist generation surfaces, thanks in part to the scientific materialism of the German Darwinist Ludwig Büchner. In short, Dostoyevsky sees that the removal of religion from the public and private consciousness leaves a vacuum, which cannot be inadequately filled by the various post-Christian imports from "Europe." But the Russian spirit is too thorough to be satisfied with ersatz religions, and thus in the 1860s it adopts an ideology of total destruction.[39]

### iv

This nihilism, sooner or later, engulfs most of the characters in the novel. Pyotr confesses that he is a nihilist, while Shigalov, the factory workers, and Mme Virginsky have nihilist ideas.[40]

Nihilism certainly implies atheism, or even becoming God, as in the case of the character Kirillov, or even becoming the "Antichrist," as in the case of Stavrogin. The atheists' tower of Babel will ascend from earth to heaven but will be built, defiantly, as a heaven on earth. But to build from the bottom up, nihilists must, of course, first bring everything from the top down.[41]

Atheism shows its face often by its total disregard of established authority. Some of the most exciting moments in Dostoyevsky's novel feature Stavrogin's outrageous breach of decorum. Stavrogin shows his contempt of the common mores by leading an old official by the nose a few steps across a room, by biting the old governor's ear on another occasion, and by kissing Liputin's wife on the mouth in Liuputin's presence. Similarly, Pyotr ridicules Lembke for collecting revolutionary pamphlets, and returns a manuscript of the renowned Karmazinov insolently crumpled.[42]

"As far as I see and am able to judge, the whole essence of the Russian revolutionary idea consists in a denial of the negation of honor," says Karmazinov. Thus the nose tweak, ear bite, and inappropriate kiss are only preliminary affronts of what is to come. Society must be destroyed before a new one can be created. But Dostoyevsky plants the nagging fear, Will this new society not pass from unlimited freedom to unlimited despotism? In pushing characterization, dialogue, and plot as far as he does to expose the danger of the "new men" and "new women," the work could be called an antirevolutionary novel, one that came the closest to prophesizing the future Soviet regime. Philip Rahv in the 1930s asserted, "History has proven Dostoievsky to be a truer prophet than Lenin." And Gary Saul Morson recently wrote, "There is so far as I know, no other nineteenth-century thinker who foresaw what totalitarianism would be like, and I know of no other literary work so dramatically vindicated by subsequent political events as [*Demons*]."[43]

In an appendix, Dostoyevsky devoted a posthumous chapter to Nikolai Stavrogin's confession to the revered monk Bishop Tikhon—a chapter that would ordinarily end Part Two but was so scandalous that it was not published during Dostoyevsky's lifetime. Stavrogin tells Tikhon of his seduction of an adolescent, Matryosha, who subsequently hanged herself. Nikolai is, of course, an atheist, but according to Tikhon, "total atheism is more respectable than worldly indifference" and "stands on the next to last upper step to the most perfect faith . . . while the indifferent one has no faith apart from a bad fear." In confessing, Stavrogin struggles to unload a terrible burden. He has printed his confession, which he intends to make public, except for one crucial page, which he withholds from Tikhon. In the remainder he recounts that he rented three apartments, two to carry out arranged love affairs and the third to live in with the Lebyadkins. One day, Matryosha was left alone with Stavrogin. He raped her and she, after shaking her tiny, clenched fist angrily at him, hanged herself. Stavrogin's guilt is indescribable. He cannot get over the suicide. He lives with pent-up fury and "conceived the notion of somehow maiming [his] life, only in as repulsive a way as possible . . . of shooting [himself]." Tikhon, sympathetic and quiet, says he is ready to forgive Stavrogin on one condition—that Stavrogin forgive himself. Tikhon then warns that the world will laugh at Stavrogin if he publishes his account: "The laughter . . . will be universal." The nihilists at the fete knew how to destroy, he says, and so does the reading public. "God will forgive your unbelief." But Stavrogin, Tikhon suggests, must fight his masochism by keeping silent. He must become a hermit's novice: "Just be a novice secretly, unapparently; it may even be done so that you live entirely in the world." Tikhon fears a worse fate for Stavrogin—that he will not forgive himself. Such a self-condemnation would be truly nihilistic. In any case, Stavrogin does not accept Tikhon's suggestion.[44]

Nihilism and atheism were apparent to the Russian intelligentsia of the 1860s—to Turgenev, Chernyshevsky, and Nechaev. Revolution, the type imported from France—1789, 1830, and 1848—arrived in Russia a generation later. By then, it had become a studied, self-conscious cult that would soon climaxe in Bolshevism.[45]

The members of the Quintet, modeled on Chernyshevsky's cell of 1848, or possibly the real Land and Liberty network, are secondary or tertiary characters, valuable as instruments. The one exception is Shatov, who no longer believes in atheistic socialism and who had the gumption to slug Stavrogin for his outrageous behavior—what he stands for—at Mme Varvara's home. This naturally raises suspicions about Shatov's loyalty. He is a marked man. He is invited to dig up the most indispensable revolutionary instrument after guns—his printing press—just as Ivanov had been ordered to dig up *his* printing press a year earlier for Nechaev. Dostoyevsky, the artist, weaves in and out between reality and fiction. The second press buried a year before can now be used to print out instructions for "the central committee," modeled on Chernyshevsky's. While Shatov is digging, Pyotr kills him with a revolver. The revolution is unkind to the Christlike ex-serf.

Before Pyotr and the Quintet settled accounts with Shatov, Shatov's ex-wife returns from abroad. She arrives in the middle of the night, abusing the driver for not knowing his way. She then reclines weakly on Shatov's bed, and it becomes clear that she is in labor, although it can hardly be with Shatov's child; indeed, this is another one of Stavrogin's outrages. Shatov runs desperately from house to house trying to find a midwife. A capable but hard-boiled matron arrives and offers to put the baby in a foundling home. But Shatov is extremely moved by the birth of this child, whose mother curses him unaccountably in her labor. The birth is described by Shatov: "Oh, it will all be without awe, without joy, with contempt, squeamish with curses, with blasphemy—this great a mystery, the appearance of a new being! Oh, she [Marya] is already cursing it now!" He vows not to send it away, and he even calls out to God. The baby dies shortly afterward.[46]

The ensuing death of Kirillov, the engineer who lives in the same house as Shatov, is more nerve-racking. For his death is not murder, but suicide. Earlier, Kirillov was angry with Pyotr because Pyotr ordered him to go to Petersburg to kill Stavrogin. Kirillov accuses Pyotr of murder and atheism. He himself believes that God is dead, and, that being the case, if he, Kirillov, takes his own life he will not only overcome pain and death but become God, man-God, because only God has the right to take life, his own life. Kirillov will be performing the highest act of will conceivable—the opposite of inventing God so as to go on living. Pyotr leaves for a few minutes. When he comes back, he finds Kirillov dead—shot through the temple.[47]

Stavrogin departs in a strange but significant way. He has been talking about going to a town beyond St. Petersburg to attend to some business, and thus conveniently skips town. He has a second-class ticket, but encounters a wealthy landowner, who invites him to join him in his compartment and play cards. Stavrogin's first-class ride symbolizes his venality and his betrayal of the revolution, for he has left behind his companions-in-arms, dead or stranded.

But is Enlightenment dead? *Demons* ends with the last trip of Stepan Trofimovich, Pyotr's father, to an unknown destination. He is ignorant of what has been going on in his son's secret society. After the festival and the fire, he sets off across the hay fields, with inappropriate footwear. He is given a ride by peasants, who are somewhat less deferential than before emancipation. Stepan learns about the true Russia through them. They want to know everything about Stepan—where he is going, why he is going there, what is his standing, merchant or gentleman. Whatever the case, he must pay them half a ruble for the ride. They take him to Spasov, a seaport, where they stop at a small inn and Stepan orders a meal. There he meets Sofya Matveyevna, a kind, gentle Christian woman who ministers to his physical and spiritual needs. He asks her to read the Bible to him—especially the passage about the Gerasene demoniac, whom Christ cured of diabolical possession. In the passage, Christ expels the demons, who then infest a herd of pigs, which then hurl themselves headlong into the sea—one of the most dramatic incidents in the New Testament. But Stepan is ill, delirious. He makes a deathbed profession of faith: "If there is God, then I am immortal! Voilà ma profession de foi"—a profession diametrically opposite to Kirillov's. Stepan is saved by helpful peasants and by the biblical account of exorcism read to him by a solicitous woman.[48]

The novel that Dostoyevsky published during his lifetime concludes with Stavrogin's returning to Skvoreshniki, Stavrogin's family home, where his mother later discovers him hanging in the loft from a silk cord. "Everything indicated premeditation and consciousness up to the last minute. Our medical men after the autopsy completely ruled out insanity." With those last words, Dostoyevsky putatively nails the lid on the revolutionary movement.[49]

Nikolai and Pyotr represent pure, rational, demonic evil. The revolutionary cycle has played itself out—from atheism to murder and suicide. Nihilism is its pure form. This small provincial town, with its succession of unheard-of crimes, and its thirteen corpses, is the victim of those who, having lost God, have lost their rationality as well.

v

Dostoyevsky is a severe critic of secularism, which, though he didn't use the word, he saw stemming from the ideas of the French Enlightenment. The Russian elite adopted the French language, constructed imitations of Versailles, patronized the French philosophes, and imitated, so they thought, the French Revolution—from the Decembrist revolt of 1825 down to the Nihilists' and Populists' terrorism of the 1860s and 1870s. Ideas transmogrified in the nineteenth century into pantheism and socialism under the influence of Hegel, Fourier, Sand, and Hugo, to mention only a few. Initially seduced by these ideas, Dostoyevsky himself made a dramatic volte-face in Siberia.

Dostoyevsky was secular himself in that he apparently went to church rarely (like Tocqueville) and loathed Roman Catholicism (unlike Tocqueville). But he repudiated modern European socialism, for he saw it as inimical to religion as well as to human liberty. He was an enthusiast of Russian Orthodoxy and of the Romanov autocracy, for Siberia amazingly reconciled him to the regime; witness the sympathetic figure of Bishop Tikhon, in his posthumous chapter of Stavrogin and of the autocracy in his "Official Correspondence" with Alexander II.[50]

During his imprisonment in Siberia, Dostoyevsky saw the evil that was penetrating Russian society—secularism or an intellectual disdain for Christianity—from the antitheological thought of the West. In place of a transcendent being, there arose in the West a worship of man himself. Socialism was this religion, Dostoyevsky believed, because it put devotion to the cause of mankind ahead of devotion to Christ.

Was Dostoyevsky, then, a true believer *and* a secularist? Yes, with qualifications. Dostoyevsky's Christianity was bound up with Russia; it was not a universal Christianity, even though he wished that it were. Rather, it was the religion of czarist Russia, a Slavophile religion. Dostoyevsky may have unwittingly subordinated the spiritual kingdom of Christ to the cause of one people, for his theocratic vision of Holy Mother Church was inseparable from his love of Holy Mother Russia.

How can this be reconciled to the transcendental characterization of good and evil in his work? The only fair answer is that, for Dostoevsky, Russian civilization *was* transcendental and the icon *was* supernatural. Whereas the Catholic West believed in the mystical body of Christ as represented by the universal church, Dostoyevsky seemed to believe in the mystical body of Russia as incorporated in the persons of the czar and the patriarch. Dostoyevsky, who condemned the temporal character of Roman Catholicism, sacralized the Orthodox clergy and the Romanov dynasty as the means of universal salvation.

In *Demons* Dostoyevsky painted a secular society in the making, which was quite different in its outcome from that of the democratic West. It would almost seem that Russia had skipped over a stage by some Trotskyite law of combined development. Russia's secularism was as extreme as its class structure, its penal system, its music, its weather, its vodka, its geography, and its autocracy.

What can be said about secularism can also often be said about socialism. Both evince a hubris, which posits that humans can live their lives individually and socially, philosophically and artistically, and scientifically and educationally, without belief in God or the practice of religion, in an immanent, autonomous, self-sufficient world order with no spiritual dimension.[51]

According to this atheism, the universe is composed only of particles of matter. There is no spiritual dimension, or if there is, it emanates from matter and human creativity (which is seen as "divine"). The only moral criteria are utility and pleasure. Neither conscience nor natural law has a real ontological existence; each is merely a human construct. There is no moral law inherent in an intelligible universe. When these beliefs have siphoned religion away from the public domain, we have secularism. The only religion left to the public sphere is the religion of the Pantheon, where man once again becomes "the measure of all things." Then the private sphere is itself invaded by the nocturnal rap on the door, and we wake up to the Gulag.

# Chapter Thirteen
# Solzhenitsyn, Communism, and the West

i

Aleksandr Solzhenitsyn[1] was born in Kislovodsk in the Caucasus Mountains in southern Russia on December 11, 1918, just a few months before Lenin unleashed the Red Army against the enemies of the nascent Bolshevik state. Aleksandr's father, Isaac, who died in a hunting accident six months before Aleksandr's birth, was a modest landowner, while his mother came from a very wealthy land-owning family. In 1919 the infant's maternal grandparents were forced off their estate by the Red Army. Aleksandr's widowed mother, who had been accustomed to comfort, was obliged to find work in Rostov as a shorthand typist. Thus the family tasted the first dregs of Russian communism. Nonetheless, Solzhenitsyn was not reared as a counterrevolutionary. He went to a public school and a high school (1926–1937), both in Rostov, where he met his first wife, Natalya (Natasha) Reshetovskaya. On the eve of Hitler's invasion of the Soviet Union (1941), he graduated with degrees in physics and mathematics from Rostov University. He thereupon enlisted in the Red Army, and rose through the ranks from private to captain by 1944, when he was awarded the Order of the Patriotic War, and saw the Soviet army advance victoriously toward Berlin in 1945.[2]

Throughout his life, Solzhenitsyn was very much a Russian patriot, but not a blind one. On one occasion, he confided some derogatory comments about Stalin in a letter to a friend, and thereby began a career of patriotic criticism. Unfortunately, his letters and other writings were picked up by a censor, and he was sentenced to eight years of incarceration and forced labor, first in the Moscow prison, Lubyanka, then in the Marfino Institute in 1947, and finally to what has become known as the Gulag, or forced labor camps. Solzhenitsyn's description of this imprisonment shows how unlike Dostoyevsky's Siberian exile it had become. Where Dostoyevsky actually came to love the czar in Siberia, Solzhenitsyn would dedicate his prison and postprison life to overthrowing Marxism-Leninism by the pen.[3]

But at this point, Solzhenitsyn was still a communist. Much later, a Gulag officer extracted an unwilling admission from him: yes, he was "a Soviet person."[4]

For the time being, he had to live down a false accusation that he had surrendered to the German army, like the hundred thousand Russian POWs who were infamously returned after the war by the Allied powers to Russia, where they "merited" the death sentence. This injustice angered Solzhenitsyn all his life.[5]

The future Nobel laureate worked as a miner, bricklayer, and foundryman in a concentration camp in Ekibastuz—an "island" in the Gulag Archipelago founded by Lenin and perfected by Stalin. A fictionalized account of this Gulag was published in 1962 in Solzhenitsyn's *One Day in the Life of Ivan Denisovich.*

Just living in the USSR was a traumatic experience. Anyone could be awakened in the middle of the night by the sharp raps of the KGB or secret police. Twenty minutes to pack all your things—no lawyers permitted. You were then dragged for interrogation to a prison such as the Lubyanka. If you were married and your wife was arrested simultaneously, the "Organs" (the Supreme Directorate of Correction Labor Camps) or police of the Gulag could torture her within earshot in order to make you confess to crimes you had never committed. Then you were taken by train through the vast archipelago—thousands and thousands of miles. As many as thirty people could be crammed into one compartment built for ten. Urine, excrement, and even corpses made it reek. Rations could be inexplicably suspended.[6]

This was the gateway to the Gulag. These were the camps that enthusiastic visitors to the USSR, such as England's acclaimed Sidney and Beatrice Webb, George Bernard Shaw, America's Eleanor Roosevelt, and later, France's Jean-Paul Sartre, were never shown when they visited and who would probably never have believed even if they had been shown the camps. (Camus wrote about them accurately in *The Rebel.*) The aim of the Gulag was to remove dissenters from the general population, submit them to grueling physical labor, and either kill them or break their will. Stalin kept a substantial percentage rolling on the tracks to this hell.

At a snow-based camp, Solzhenitsyn found 323 men crammed into a 20-man cell. Forty thousand miserable human beings inhabited one prison, where 200-watt bulbs were lit up at night to ensure order. Corpses that were too tardily removed made life unbearable. Stool pigeons listened to their conversations to report them to the guards. Common criminals were mixed with the political prisoners to lower morale. Independent scholars put the death toll in the camps at seven million between 1933 and 1940 alone.[7]

The most horrendous practice of the Gulag was the treatment of women, who were paraded naked upon arrival so that the guards and Organs, or "trustees," could choose which ones they wanted. Chastity, Solzhenitsyn argued, was not feasible for separated couples, and he urged his wife to get a divorce, which she did in 1951. Women who did bear children in the Gulag could not christen them or raise them decently. "They grow up hooligans and petty thieves and the girls . . . run free and loose." From girlhood, women worked eight-hour days on the conveyor belts—the kind of work with which Marx explained capitalism.[8]

Some women were exceptionally courageous in face of the Gulag. For instance, one Vera Korneyeva belonged to a group of seventeen who refused to be cowed by the Organs; she opposed them:

> Why do you persecute your best citizens [religious believers]? They represent your most precious material after all, believers don't need to be watched, they do not steal, and they do not shirk . . . Everything in this country is falling apart. Why do you spit in the hearts of your best people? Separate church and state properly and do not touch the church; you will not lose a thing thereby. Are you materialists? In that case, put your faith in education—in the possibility that it will, as they say, disperse religious faith. But why arrest people?[9]

These were bright moments of truth, when the ongoing repression was overpowered by the strength of individual conviction.[10] The personal testimony of victims of totalitarianism is gut-wrenching, as are the sheer number of deaths. Solzhenitsyn estimates the extermination of the kulaks, the purges of the thirties, and the deaths in the Gulag at sixty-six million—all the victims of the Soviet regime. He contrasts these numbers with the 16,000 criminals or prisoners shot under all the czars. The czarist regime, he argues, was far more lenient, less chaotic, and far freer than the Stalinist regime. It was Lenin, he reveals, not Stalin, who coined the term "concentration camp" and invented the reality that Hitler later borrowed. Solzhenitsyn irresistibly indulges in black humor when commenting on the period of the civil war. He writes,

> It turned out that capital punishment had been . . . abolished! What a fix! It just couldn't be! What had happened? . . . [The Cheka, without capital punishment?] But had it been extended to the tribunals by the Council of the People's Commissars? Not yet. Krylenko cheered up. And he continued to demand execution by shooting.[11]

Worse than death was the hopelessness. A Captain Medvedev gloated over some prisoners' condition: "You have lost the game! There is never going to be any return to freedom for any of you, and don't dare to hope there will be." One day of forced labor after another, gruel to eat, no family or friends. "Shall we sum up the whole of Russia in a single phrase?" Solzhenitsyn asks. "It is the land of smothered opportunities." There was no worse regime in history: "Not even the regime of its pupil, Hitler, which at that time blinded Western eyes to all else."[12]

Solzhenitsyn attributes the Gulag atrocities to ideology: wholesale, collective obedience to a secular worldview that subjects all thought and action to one goal—in this case, the total global victory of communism. Lenin said, "Morality is what is good for the state." That summed it all up. Any action could be justified, including the wholesale liquidation of certain social classes, which was as monstrous as Hitler's liquidation of certain races.

Ideology explains extraordinary evil. Shakespeare approached something close to pure evil with Iago and Macbeth, but, Solzhenitsyn remarked, "Shakespeare's evil-doers stopped short at a dozen corpses, because they had no *ideology*." Ideology "gives evil-doing its long sought justification . . . That is the social theory which helps to make [a man's] acts seem good instead of bad in his own and others' eyes." Thanks to ideology, the twentieth century was fated to experience evil-doing in the millions.

Solzhenitsyn asks where this "ideology" came from. He bypassed all the nineteenth-century Russian revolutionaries, such as Chernyshevsky and Nechaev, and fingered the Englightenment. "The murky whirling of *Progressive Ideology* swept in on us from the West at the end of the last century," he explained in January 1974, echoing Dostoyevsky. It was "dinned into our heads by the dreamers of the Enlightenment," whose promise of "*endless, infinite* progress" was unreliable. Any peasant would have understood that this was impossible: "For two centuries Europe has been prating about equality—but how very different we all are! How unlike are the furrows life leaves on our souls." Although East and West are very different, they have been joined by a common disastrous ideology.[13]

"The French and October revolutions have this profound similarity: they were both *ideological;* and they did not annihilate people haphazardly but on the grounds of *ideology*."[14] "Ideology" was a social-scientific term, which was given its modern meaning by Marx, who saw the individual as a product of his social and economic circumstances. Solzhenitsyn, who observed gross collective evil on an unprecedented scale, believed that it was an evil system of thought used by Marx and Lenin to justify atrocities for the sake of a supposedly noble end. Beyond this, Solzhenitsyn did not go.

Communists used ideology as a substitute for thought and law: "There is no need to clarify whether the defendant is guilty or not guilty," Solzhenitsyn quotes Drylenko, Stalin's state prosecutor in the thirties. "The concept of guilt is an old bourgeois concept, which has now been uprooted." The classless society demands trials without conflict—for the only real conflict is class conflict. Therefore, the perfect communist trial is one in which the accused is also the accuser.

There was no recourse, therefore, to law or lawyers in these famous show trials of the 1930s, in which high-ranking Party members finally accused themselves of crimes they had never committed, because ideology dictated that, given their class and upbringing, they must have committed them. No natural (or unnatural) rights existed in this system. This condition was improved upon by the Bolshevik prison keepers, who said to the peasants, "It's your fatherland—you defend it, you dung-eaters! *The Proletariate has no fatherland!*"[15]

A denunciation, in this system, "is not a denunciation but a *help* to the person denounced. If the Organs wish it, the recruited person must also wish for one and the same thing: our country's successful advance to socialism." This identity of the will of the *zek* (prisoner) and the Soviets is reminiscent of Rousseau: The individual dedicates his whole self to the general will and still obeys only himself. Freedom and obedience are identical, as in the life of a cloistered monk. But ideology justifies all the strange and oppressive practices employed to suppress a prisoner's consciousness or to "elevate" him by brainwashing in mental asylums.[16]

Writing shortly after his exile abroad, Solzhenitsyn claimed that ideology was dead but still held out its "claws." "The State system is founded on ideology . . . all we have to do is to renounce the ideology, stop supporting it, and the State system will collapse by itself—it has nothing to hold on to but ideology."[17]

Perhaps the most interesting and impressive observation Solzhenitsyn made is how communist ideology could be used to explain good and evil. He says, shockingly, that had circumstances been otherwise, he could have been a member of the Organs, gone to executioners' school, and become an agent of Stalin's secret police. Solzhenitsyn argues that every man has good and evil in his heart, and that the line distinguishing him from others is constantly shifting, not so much as a result of his free will, but because circumstances can change him into "a totally different human being."[18]

This idea seems to undermine the clear language of free choice in his three volumes of *The Gulag Archipelago*. Its merciful forgiveness of the Soviets can be read as paraphrasing in the vernacular what Christ had said of his executioners: "These poor bastards did not know what they were doing. They were entrapped by circumstances."

This awareness of Solzhenitsyn is not so much a statement of Pavlovian psychology (that man is what he is conditioned to be) or a denial of free will (which Kant considered prerequisite to moral behavior), but a reprieve from his unrelenting incrimination, an act of charitable understanding. He does not want to be self-righteous. He himself, after all, was once "a Soviet person."

Yes, *The Gulag* is a political exposé, but a compassionate, moral, and religious exposé. Solzhenitsyn could sympathize with the millions of communists sucked into the evil vortex by circumstances. But his sympathy stopped short of condoning the originators of the ideology: Marx, Lenin, and Stalin. He was also quite unsympathetic to those after 1953 who attempted to save communism by discarding Stalinism. The ideology, not the men, was utterly evil.

*One Day in the Life of Ivan Denisovich* (1962) was Solzhenitsyn's first book, a brief novella about a *zek* (Ivan Denisovich Shukov) who is serving a ten-year sentence in the Gulag. Every minute of his twenty-four-hour day is detailed and conveys a sense of drudging purposelessness found earlier in Sartre's *Nausea* (1938) and Camus' *Myth of Sisyphus* (1942) and *The Plague* (1947). Solzhenitsyn excels in a revitalizing realism, turning it against the Soviet system. In the book, Ivan's life is depicted as a shuffle for survival. There is no hint of religious or philosophical self-sacrifice—no voluntary acceptance of suffering. The only search for broader meaning comes with the character Alyosha, a Baptist, who believes that God can move mountains, whereas Ivan, the protagonist, believes that his petitions to God get rejected exactly like those in the camp petition box. Meanwhile, Zhukov says he is not against God, but does not believe in paradise or hell.

We are treated to a depressing account of a 200-bed dormitory: rude awakenings; fights over gruel, sausages, or tobacco; guards constantly pitted against prisoners whom they called "savages"; reprimands over the endless work. *Ivan* is not autobiographical, but it draws from the author's prison experience. However, many of the details are changed to heighten the effect—for example, Shukov's sentence is twenty-five years instead of the author's eight.

## ii

When Solzhenitsyn wrote *Ivan,* he had been out of the Gulag for six years and was in exile in Kok Terek in southern Kazakhstan, where he returned to teaching school. He and his ex-wife remarried. He lived with her until he had two sons by Natalya Svetlova, whom he married after his second divorce from Natasha in 1970.[19]

In 1957 he moved to Ryazan (near Moscow), where he taught mathematics. When he came up against constant procrastination by the publishers of the journal *Novy Mir,* in which *Ivan* would be published, he persisted, meeting frequently with its director, Aleksandr Tvardovsky, who became a close friend. Tvardovsky constantly asked for the manuscripts that Solzhenitsyn had submitted for publication. The publishers, it seems, did not want to publish the book themselves or let anyone else do so. But they never said that. Finally the matter was settled. But before the publication of *Ivan,* objections arose over a passage about God. This passage was "artistically ineffectual, it was ideologically incorrect, it was too long and it only spoiled a good story." There was another objectionable conversation: "I crossed myself, and said to God 'Thou art there in heaven after all, O Creator. Thy patience is long, but thy blows are heavy.'" Clearly, this religiosity flew in the face of Soviet secularism, which banned all but pejorative mention of religion. The parenthetical "after all" could well refer to Solzhenitsyn's personal discovery of God after years as an atheist. He had referred to the "Creator" in a series of prose poems called *Miniatures,* which provoked Tvardovsky to ask about "The Creator, and with a capital C. What is all this?"[20]

Solzhenitsyn refused to change the manuscript he had given to *Novy Mir.* "I've waited ten years and I can wait another ten . . . Give me back my manuscript and I'll be on my way," he said. But they held onto it and continued to place roadblocks to its publication. The Politburo obviously had a finger in the works, and Tvardovsky was afraid to publish anything that might endanger his position. But he was also a well-known poet, whom Solzhenitsyn and the literati respected. "I owed my rise to this one man: perhaps my survival also depended on him alone," Solzhenitsyn later wrote. Tvardovsky did not betray Solzhenitsyn; he pleaded with him. Solzhenitsyn knew how to wait; he had waited in prison and would continue to wait as a "free man."[21]

*Ivan* created a sensation, selling out 95,000 copies of *Novy Mir* in one November day of 1962. Solzhenitsyn became instantly famous, gaining a worldwide reputation. Interest in what really went on behind the Iron Curtain accounted for its success among Soviet citizens as well as foreigners. Hundreds of ex-*zeks* wrote letters to him, offering him their own lurid tales of the Gulag experiences. These later became the basis of his much fuller account in the three-volume *Gulag*, a work of "literary investigation." But *Ivan* and a few short stories such as "Matronya's House" were his only publications in the Soviet Union before his exile in 1974. His eminently political novel *First Circle,* about a scientific institute under Stalin, was confiscated by the KGB but then published abroad in 1968, as was *Cancer Ward.* The success of *Ivan* gave him enviable access to the international press.[22]

Before his exile, the KGB tailed him, intercepted his mail, broke into his home, and seized his manuscripts. Their surveillance was ordered from the pinnacle of government, the Politburo, which included Brezhnev and Andropov after Khrushchev's fall in 1964. (Documents of this surveillance were made available after the fall of the USSR in 1991.) Chiefly involved were the Council of Ministers and the Committee for State Security of the USSR. Their minutes reveal concern, annoyance, and alarm at the anti-communist, "anti-social, anti-Soviet" propaganda in Solzhenitsyn's *First Circle* and *Cancer Ward.* What is most impressive about this high-level surveillance is how seriously these apparatchiks took this math teacher from Ryazan. They concluded repeatedly that more needed to be done on the "ideological" front to strengthen the conviction of artists and writers concerning the rectitude of the Soviet cause. They were keenly aware of the power of public opinion at home and abroad, especially in France and Yugoslavia. And they were right: it would be this force that would save Solzhenitsyn, for harass him as they would, the authorities could not "punish" him as they would have liked, even though he was recorded as saying that the Soviet Communist Party was "bankrupt" and that "the Archipelago will murder them. It will be their devastation."[23]

In the end, *Novy Mir* refused to publish *First Circle* because of its anti-communist, anti-Soviet passages: "Can't the author think a bit more kindly of people and of life? . . . He simply does not love the people." This accusation was frequently thrown at Solzhenitsyn, not only in the USSR but in the United States as well. No, he did not mince criticism. But then, there was nothing redeemable about the Gulag. *Novy Mir,* under Tvardovsky's successor, wanted to look on the Soviet society "realistically," which meant benevolently. Unearthing the millions of skeletons of Stalinism threatened to widen the crevice, which Solzhenitsyn prided himself in broadening into a crater.[24]

Simultaneous to his fruitless negotiations with *Novy Mir,* Solzhenitsyn was expelled from the Soviet Writers Union at Ryazan, where he was cited for "hostile bourgeois propaganda for a campaign of slander against our country" and of "fan[ning] the flames of anti-Soviet sensationalism around his name." He did not take this lightly, and responded at a meeting with those writers in November 1969: "Blow the dust off the clock. Your watches are behind the times. Throw open the heavy curtains which are so dear to you—you do not even suspect that the day has already dawned outside . . . OPENNESS, honest and complete OPENNESS—that is the first condition of health in all societies, including our own." Solzhenitsyn was among the first to call for glasnost, although not the Gorbachev version.[25]

Convinced that the "Soviet regime could certainly have been breached only by literature," and that he could never publish legally in the Soviet Union, he created an underground self-publishing network, which was called samizdat, and which is described dramatically in *The Oak and the Calf* (1975). The most important "links," or agents, were described as "Invisible Allies"—a kind of literary corps of counterespionage, consisting mostly of women who would hide, copy, and pass on manuscripts for Solzhenitsyn. Sometimes they buried texts. Other times Solzhenitsyn made microfilms after photographing them in his home. He then took a hardcover book, pealed off the binding, inserted the film inside each leaf, and sewed it back together again. The KGB never found them. They were handed over to Westerners at airports, from where they were smuggled past Soviet customs. Eventually, these copies were published in the West on a "signal" from Solzhenitsyn.[26]

After Khrushchev's fall from power in 1964, Solzhenitsyn was gratified to learn that some of his works were being printed by "Grani," the Russian émigré press abroad. He managed to get three copies of *First Circle* from *Novy Mir* and deposited them and some other manuscripts with a friend named Teush in Moscow. The samizdat was working. But all the while, he continued to be tailed and his phone was bugged. Tsezarevna Chukovskaya (Lyusha), "the Paganini of the typewriter," typed five hefty volumes in three years, mostly without compensation, while holding an office job. She was soon assaulted in her apartment house, and subsequently targeted and seriously injured by a KGB truck. She still managed to have her copy of *Cancer Ward* smuggled out of the country.[27]

Elizaveta Voronyanskaya, another samizdat participant, who held the only complete manuscript of the *Gulag,* was arrested in 1973 and interrogated by the KGB for five days; she finally broke down and revealed the location of the whole manuscript. After a cardiac arrest, she was hospitalized.

Upon her release, feeling like a "Judas," she hanged herself "in that dark, fetid Doskoyevskian hallway" of her home.[28]

Solzhenitsyn may have felt remorse and sorrow at the fate of these "companions in arms," but he also seemed to feel like an officer who cannot grieve over fallen bodies but must continue the fight: "Not for a single hour, not for a minute, was I downhearted on this occasion. I was sorry for the poor, rash woman whose impulse—to preserve the book in case I could not—had brought disaster upon it, upon herself and upon many others. But I had enough experience of such sharp bends in the road to know from the prickling of my scalp that God's hand was in it!" His only wish was "to live up to the hopes that reading Russia places in [me]."

In the meantime, under continued fire and surveillance, he acted with the singleness of purpose of an intelligence officer, mending the tears in his network as quickly as possible and ordering documents to be burned and buried whenever he thought the KGB was on the scent. He was acting the part of a rational, mathematical strategist. He vowed with Natalya, his wife, not to succumb to blackmail but to continue their project even should his children be taken hostage. Eschewing his professions of nonviolence, he now spoke continually in metaphors of plans to "detonate" the "rocket" and "bombshell." He waited for the right moment to launch the little capsule of microfilm containing the *Gulag* with the instruction "print." These preoccupations took precedence over his family.[29]

### iii

Solzhenitsyn then went "public" by accepting speaking engagements in Moscow. "This was perhaps the first time, the very first time in my life, that I felt myself, saw myself, making history . . . Almost every sally scorched the air like gunpowder!" But almost immediately, he had to go into hiding for seventy-three days, from December 1966 to February 1967, during which he typed 1,500 pages. "I produced my final draft of the *Gulag Archipelago*," he exulted. "It was not I who did it—mine was merely the hand that moved across the page!" Solzhenitsyn no longer saw himself as the ex-*zek*, the cranky dissident, or the pleading author, but as a semireckless, semi-inspired man, ready to take enormous risks. He must shout louder—"a shout in the mountains has been known to start an avalanche." The avalanche would be public opinion growing inexorably, eventually overwhelming the Soviets.[30]

Thanks to the publication of *Cancer Ward* and *First Circle* in the West in 1968, Solzhenitsyn became nearly untouchable. Should anything happen to him, it would provoke an international uproar and undermine the détente between the USSR and the United States that was then under way. Tvardovsky's final words to him had been that he must compromise with the Soviet system: "*You hold nothing sacred . . .* You must make some concessions to the Soviet regime . . . you can't afford not to—you can't fight a howitzer with a peashooter." Solzhenitsyn countered, "We have a howitzer of our own." And indeed he did: public opinion.[31]

Solzhenitsyn wanted to publish in Russia only the complete and authentic version of the *Gulag*, but he realized the impossibility of it: "I had been allowed to get away with a lot. But I would not get away with *Gulag*." But the taste of victory, even at the price of his life, became irresistible. The fate of *Novy Mir,* the Politburo, and millions of Russian readers as well as Western opponents of appeasement was in his hands. On the third day of the Feast of the Trinity, in June 1968, he received a coded message that the *Gulag* had reached America and was in the possession of Olga and Henry Carlisle at Harper and Row in New York, who were to prepare a translation by 1970, when he would give the signal for publication.[32]

"All the paths I had chosen," Solzhenitsyn wrote, "all the moves I had made with my books and my letters, seemed now to have been devised not by my poor human brain, and sheltered by a shield which was certainly not mine." Over and over, Solzhenitsyn stressed that this huge undertaking was being guided by Providence. How else explain its success against such enormous odds?[33]

The invasion of Czechoslovakia in August 1968 increased the nervous tension. He listened to the Voice of America on the radio and whatever else that was not being jammed. Tanks were rolling past his dacha in the outskirts of Ryazan, where he was writing his novel. "How uninventive they are, how underendowed with thinking power." But they "entered" Czechoslovakia and "crushed it. Which means, to the twentieth-century mind, that they were *right.*" An attempt to draw up a letter of protest signed by prominent Soviet figures—including the musicians Rostroprovich and Shostakovich—got nowhere. Solzhenitsyn was still a Soviet citizen, but no longer a "Soviet person." "I am ashamed to be Soviet!" he repeated. A few weeks later, he installed himself in Rostoprovich's elegant house outside Moscow, where he was visited by "gasmen" and "electricians" and was "tailed" when he went for a walk. His metaphors of defense grew. He now had "a rocket so powerful that [he] . . . was almost sorry not to have to launch it."[34]

Solzhenitsyn had enjoyed his influence as the author of *Ivan*. He received even more leverage when he won the Nobel Prize in 1970 (for works published abroad as well as for *Ivan Denisovich*). And he did not want to imitate Pasternak, one of three who preceded him as Russian Nobel laureates, or Jean-Paul Sartre, both of whom refused to go to Stockholm. The danger was that if he should go, he might never return. Nothing better illustrates his love of Russia than his unwillingness to take that chance. Instead, he negotiated, as he had with the publisher Tvardovsky. Couldn't a ceremony be held in the Swedish Embassy in Moscow, where he could give his speech? This idea was entertained for a while but ultimately dismissed by the Swedish Embassy. The award and speech were put off until further notice.[35]

Solzhenitsyn's Nobel acceptance speech was mailed to Sweden, rather than delivered orally in Moscow. In it he lauded the power of literature, through its beauty and truth, for its ability to expose and disarm a world of deceit and violence. It was as much addressed to the world as to the USSR; its main thrust was against materialism, terrorism, and appeasement, all of which dominated the twentieth century. The persecution of Christians, the assault upon the higher cultural values, and the Chinese Cultural Revolution all must be decried without compromise. This was the calling of literature that had no walls.[36]

The Nobel was "just what I needed to make my great break-through when the time came," he said. But the lecture "raised not an eyebrow among our masters . . . though crystal clear, [it] was written in general terms, without a single proper name . . . The Nobeliana was over, and the explosion, the main battle, was still indefinitely postponed."[37]

The camp experience had developed Solzhenitsyn's calculated audacity. "It's impossible for anyone trained in the camps to . . . lose . . . self-control." Hence his patience, his chesslike strategy in dealing with the harassing *Novy Mir,* and the even worse KGB. The Soviets had inadvertently trained him as a prizefighter in this war of nerves. And he insisted that he would win: "Strength and steadfastness are the only things these people fear," he stated. He appealed to the rights and duties of the Soviet citizen. "That's the way to talk to them!" Over the following months, Solzhenitsyn continued his brinkmanship. In the last volume of the *Gulag,* he wrote, "No other regime on earth could compare with it either in the number of those it had done to death, in hardiness, in the range of its ambitions, in its thoroughgoing and unmitigated totalitarianism—no, not even the regime of its pupil Hitler."[38]

Solzhenitsyn's 1972 "public" but little publicized *Lenten Letter to Pimen, Patriarch of All Russia* is unique and revelatory. To start, he notes that the vast majority of churches have been destroyed, abandoned, or desecrated.

He then goes on to say that the state of the Russian church during the chairmanship of Khrushchev and after was worse than it had been under Stalin. He upbraids the patriarch for being so timid and submissive to the state. Why, he asks, "should I have to produce my passport when I come to church to christen my son?" Such registration compels courageous parents "to inform against themselves to the state." More obstacles were placed before parents who wished their children to receive religious instruction and the sacraments. Priests were not allowed to leave their churches, he went on, and the contributions of parishioners for the poor were taken by the Committee on Religious Affairs, leaving beggars unsuccored on the church doorsteps. A church directed "dictatorially by atheists" had not been seen in two millennia. From the catacombs to the Gulag was Solzhenitsyn's idea, if not his words—in summation, a ray of hope: The fetters of secularism, the subjection of the spiritual to a hostile and persecuting temporal power, could, he believed, be broken as they had been back in the ancient world. "Ask our Lord," he begged the patriarch. "What other aim can there be for your service amongst the people, who have almost lost both the spirit of Christianity and the very semblance of Christians?"[39]

In December 1973, the first volume of the Russian version of the *Gulag* was published by the YMCA Press in Paris. It took the Politburo and the KGB six weeks to appraise the damage and react. Their options were limited: execution, imprisonment, psychiatric hospitalization—but all were likely to cause worldwide uproar and threaten détente. These were the years when the eyes of the world were focused on the dissidents inside the USSR. No, exile of this giant dissident might be their only recourse. It would be a grand gesture, a defiance of the author's charges of their inhumanity.

But Solzhenitsyn had won: "The calf butted the oak: a futile enterprise, you might think. The oak has not fallen—but isn't it beginning to give just a little? As for the calf, his forehead is intact, as are his budding horns." The metaphors of bombs and explosives were discarded. As things reached a peak in early 1974 and reports started circulating that he was going to be exiled, he appeared to revert to his earlier truculent command at *Novy Mir*: "Give me back my manuscript."[40]

Solzhenitsyn's arrest on February 12, 1974, was sudden and unexpected. He was given a few minutes to get some things together, without being told where he was going. Neither his wife nor his children could accompany him. He gave Natalya some instructions concerning the "Solzhenitsyn archive" and the samizdat. He was then hauled off to the Lefortovo Prison, where he was told by a guard that he was going to be executed. Yet another lie. In the morning, he was put in a car with KGB officers and taken to the

Sheremetyovo Airport, where he took a scheduled flight, delayed three hours for the purpose, and landed in Frankfurt, where he was met by German officials and taken to Heinrich Böll's house.[41]

### iv

Solzhenitsyn did not give up his fight against communism when he left Russia. Rather, he opened up a new front—a battle against what he deemed Western illusions and shortcomings. Indeed he saw the West on the verge of collapse in spite of its extraordinary nuclear arsenal. The West could be subverted internally by its own lack of character—its materialism and softness, its spoiled children, its promiscuous youth, and its penchant for easy solutions such as détente. In short, what Solzhenitsyn set out to do was to "warn the West." This was not immediately understood or appreciated. How could the West, whose free press published his work intact, and which enshrined freedom of religion, speech, movement, private property, the right of association, the emancipation of women, and the equality of races, be accused of not fostering things favorable to the development of the spirit? Solzhenitsyn had plenty of opportunity to answer these questions to audiences that were somewhat perplexed as to how he could bite the hand that offered him a home.

His problems in Europe, and later in America, were quite different from those in the USSR. No longer did he need to twist arms to get something published. Dozens of reporters congregated outside his domicile, wherever that was. (He had to retreat to Cavendish, Vermont, where he lived with his family from 1976 until 1994.) It was a matter of choosing what to say, to whom to say it, and getting it printed correctly in English.

"Written some time before the seizure of *The Gulag Archipelago* by the KGB," that is, in 1973, Solzhenitsyn's "Letter to the Soviet Leaders" was published in English in 1974. Headed on a disastrous course since 1917, the USSR was responsible, according to Solzhenitsyn, for the deaths of 66 million human beings, outside of the two world wars. A third world war, he warned, was impending; China would invade Russia with unlimited ground forces. (Solzhenitsyn did not see nuclear weapons as the cause of twentieth-century woes, nor their buildup as the solution.) But the destruction of the USSR was no less likely. Elsewhere, he made it clear that he was not an advocate of physical revolution either. When he spoke of standing firm, he certainly envisaged a risk, but his life had been full of risks and he did not flinch from applying lessons of his individual experience to states as a whole. It was less a question of careful strategy, such as that of former ambassador George Kennan or Secretary of State Henry Kissinger, than a moral problem. Ideology was responsible for the disaster of Russia's Gulag, its collective farms, and the serflike status of ordinary Russian citizens. The worship of "'progress' [has been] dinned into our heads by the dreamers of the Enlightenment," he insisted. Turgot's and Condorcet's dream of indefinite perfectibility had proven a nightmare.[42]

Instead of a life on the treadmill of production, Solzhenitsyn advocated zero growth and a Pascalian retreat to the inner self to cultivate man's neglected spirituality. The Soviets had to go back and start again along the right course, not just "blindly" copy "Western civilization." Russians must return to the land and develop the great northeast, Siberia. Many of his proposals resemble those of the European Greens.[43]

Solzhenitsyn believed that ideology was what led Russia astray, in particular the ideology of the Enlightenment that shut off the ascent to God. On the other hand, he saw Christianity as "the only living spiritual force capable of undertaking the spiritual healing of Russia." If they were to give up Marxism-Leninism, the Soviet leaders could continue to govern. Solzhenitsyn was not even asking that they be overthrown, but appealed for their conversion. Only someone with such extraordinary prestige could dare to speak so boldly.[44]

Solzhenitsyn's courageous statements continued to pique the interest of, and shake, American public opinion. Even in seclusion, no intellectual of the twentieth century—not Sartre, not Brecht, not Bertrand Russell—had been so much the center of attention.

In an article in the 1974 collection *From under the Rubble,* Solzhenitsyn used his earlier image, "refusing to participate personally in the lie." Marxism was the "fetid root of present-day Soviet life," he averred. His famous Russian colleague, the physicist Sakharov, who was still behind the Iron Curtain, believed that the solution lay in the mere recognition of human rights and the creation of a multiparty system. This, indeed, was what the United States sought for the USSR. But for Solzhenitsyn, the issue went far deeper; it involved stepping outside the regime entirely. Sakharov and America echoed the complete secularist's view of the problem. Morality and religion were not at issue for them, but they were the only issue for Solzhenitsyn. A purely political solution would be to "take part in the lie."[45]

A year later, despite the swarm of reporters still buzzing around him wherever he was, despite the chain-link enclosing his 50-acre vermont estate (to fend off a still active KGB), Solzhenitsyn answered the puzzling question of why the USSR had let him go. He believed that they thought he would be ignored in the West—and he thought they might have been right. Some

Western critics had begun to dismiss him as someone who simply wants a "return to Orthodoxy," and from that perception, others deduced a return to anti-Semitism. His comparison of Russian communism's casualties with the German annihilation of the Jews was doubtless felt by some as minimizing the horror of the Holocaust. In fact Solzhenitsyn, unlike Dostoyevsky, rarely spoke of the Jews, and then not as an anti-Semite. Indeed, his second wife, Natalia, was Jewish.[46]

But Solzhenitsyn's real concern was that the West's pursuing the secularist course meant ignoring the danger of the East. The West was abandoning the cause of Russian freedom, of people like himself, but it was also abdicating its own freedom. America, he thought, was traversing the course that Russia had traversed in the early twentieth century, one that would eventuate in a Marxist revolution of its own. "The question is not how the Soviet Union will find a way out of totalitarianism but how the West will be able to avoid the same fate," he said. Again he put his interlocutors on the defensive.[47]

Solzhenitsyn disliked détente because he believed it had facilitated the communist bloc's takeover of dozens of countries over the years, most recently Portugal, Angola, and Vietnam. The communist bloc would benefit from any slowing down of the arms race. The West's show of resolve in Korea, Berlin, and Cuba, on the other hand, proved that the Soviets always backed down when they came up against firmness.

What had all this to do with secularism? Invited by Jesse Helms, Solzhenitsyn first declined but then agreed to address Congress, and later the AFL-CIO (American Federation of Labor and Congress of Industrial Organizations), hosted by George Meany. In his speech to Congress, he claimed that the West had been too weak toward the USSR. He alluded to a festering grievance: the return of Russian POWs to the Soviets after the Second World War. Subsequently, the United States had supplied the USSR with necessities through entrepreneurs such as the leftist multimillionaire Armand Hammer. Trade was the great anesthetic of the Cold War; it ignored the fact that the USSR was guilty of murdering millions of innocent people and would continue to devour them unless faced with firmness.[48]

Helms proposed unsuccessfully that the Senate grant Solzhenitsyn honorary U.S. citizenship, previously given only to Churchill. The further suggestion that President Gerald Ford receive Solzhenitsyn at the White House was opposed by Henry Kissinger (who later apologized) on the grounds that it would endanger détente.[49]

Throughout, Solzhenitsyn thought the real core of the problem was the idea, held by most journalists, that morality had no place in politics. He was, that is to say, challenging pure Machiavellianism, which was so influential in European politics from the days of the Renaissance to those of Hitler. The conflict between East and West was not, Solzhenitsyn felt, a matter of counting warheads but of exposing "this deadening killing ideology." Again and again he said annoyingly, if not antagonistically, that he would be surprised by the collapse of the USSR, but not by "the sudden and imminent fall of the West." Nuclear détente was less important than "ideological détente." As long as the West held fast to the Enlightenment (Benthamite) hedonism, it would surely fall. "We have become hopelessly enmeshed in our slavish worship of all that is pleasant," he contended, "all that is comfortable, all that is material—we worship things, we worship products."[50]

## V

In June 1978, three years after he had arrived in America, Solzhenitsyn was invited to deliver the commencement address at Harvard University. What a podium! Thirty years earlier, George Marshall had delivered the famous plan, subsequently named after him, of massive foreign aid to reconstruct a Europe devastated by the Second World War. The Harvard trustees and faculty were well aware of Solzhenitsyn's criticism of the American scene, which is perhaps why they had invited him. They, like most Americans, had indulged, even wallowed recently, in guilt over the Vietnam War. But none of this was enough to prepare them for the "root and branch" tirade they were about to hear. Speaking the "bitter truth," he claimed that America was mired in the pleasures of the body rather than those of the soul, in material possessions and physical comforts rather than in moral values and spiritual goods. The public, lacking in courage, was depressed; governments were constantly hemmed in by egregious human rights, which constricted appropriate action. As for the press, he continued, it was "hasty, immature, superficial and misleading." King of fashion, he went on relentlessly, it wielded power that was never subjected to a vote. In consequence, an obscure academic would never find a forum unless his or her opinion conformed to the reigning viewpoint. Discouraged by this state of affairs, some became attracted to socialism and to the USSR, whose suffering had led to a "spiritual development of such intensity that the Western system in its present state of spiritual exhaustion does not look attractive." And finally, he thundered, America's pop music and advertising, which symbolized the general "decline of the arts," made America a "threatened or perishing society."[51]

Did the secular question arise openly? Solzhenitsyn omitted the word, but he dwelled on the substance. Not so George Kennan, who said that in foreign affairs, "we cannot apply moral criteria to politics." But morals *do* matter in politics, Solzhenitsyn insisted, flatly contradicting four centuries of Western Machiavellianism. America *is* guilty of genocide in Vietnam not for using napalm, but for abandoning millions of South Vietnamese to their North Vietnamese executioners after the peace.[52]

From where did all these distinctive ideas arise? Solzhenitsyn expanded upon his critique of the Enlightenment and pointed to its origins in "rationalistic humanism or humanistic autonomy," or the "anthropocentricity" of the Renaissance. This misconception opened the way to blame evil not on man, who was naturally good, but on society, which would therefore have to be completely reformed. In so saying, Solzhenitsyn of course risked alienating most liberal arts colleges, whose professors, for more than a century, found liberation (from obscurantism, from the church) in those two intellectual movements—the Renaissance and the Enlightenment—as well as in the social sciences to which they gave birth.[53]

Thereupon, ignoring the Anglo-Saxon panacea of "limited government," Solzhenitsyn laid the blame on rationalistic humanism, or humanistic autonomy from any higher force, and in its place "worshiping man and his material needs." "Man's sense of responsibility to God and society has grown dimmer and dimmer . . . All the celebrated technological achievements of progress, including the conquest of outer space, do not redeem the twentieth-century's moral poverty, which no one could have imagined as late as the nineteenth century." America is headed toward Marx's "social humanism"—blaming evil on social structures, rather than on man. The circle closes, for communism also lives in order to find this "earthly happiness." This, then, is "the calamity of an autonomous, irreligious humanistic consciousness," whereby "we have lost the concept of a Supreme Complete Entity."[54]

This Supreme Complete Entity would be best understood as the summum bonum—man's ultimate good. This being serves as the rival and check to the totalitarian dictator's unlimited desire for power, Solzhenitsyn went on. Man must acknowledge his limits before this Supreme Entity. The prospects are bleak—not limitless, as the modernists proclaim. If war does not destroy us first, we must reform ourselves. The transition man undergoes today is greater than that from the Middle Ages to the modern world. "No one on earth has any other way left but—upward," he concluded.[55]

Any address is interesting for its omissions as well as for what it says. While Solzhenitsyn makes one or two references to Christianity, his theological framework in this speech is more theistic than Christian—perhaps in view of his pluralistic audience. His critique of rights could have referred to the legalization of abortion and homosexuality, but he does not mention those. He does cite pornography as an abuse of freedom of the press and of the individual. Absent from the speech are the words "secularism" and "atheism," which he speaks of frequently elsewhere. He did humor his audience to some degree, but on the whole, the speech came as a thunderbolt even to those familiar with his recent pronouncements.

Solzhenitsyn did not go so far as to tag secularism for reducing discord at the price of suppressing dogma. The secular movement, as opposed to the sectarian movement, seeks to establish a common denominator by suppressing all public expression of beliefs that are not held in common. After so much reduction, the only consensus left is that of material well-being. The factor not mentioned in Solzhenitsyn's speech is the fragmenting of Christianity into so many voices that it left with no unified public expression. The outcome, Marx showed, was the atheistic state (a condition that is not necessarily invalidated by congressional chaplains, coinage inscriptions, and Gallup polls of religious practice).

The reaction among intellectuals and journalists to the Harvard address was mixed—more critical than laudatory. Many called it a great, perhaps the greatest, public statement about the American character and society since the Second World War. Virtually all of the eighteen commentators, collectively published by Washington's Ethics and Policy Center, spoke of Solzhenitsyn not only as a great man but as a giant who had taken on the monolithic Soviet Union single-handedly—and succeeded. As for the rest, almost everyone admitted that American society was very materialistic, but hurriedly added that this was not news. They seemed to grant grudgingly that Russians may have become more spiritual through their suffering under totalitarianism than had Americans through abundance. But there is a sense of hurt pride among some of these commentators—hurt, as the columnist Mary McGrory put it, because Solzhenitsyn "doesn't love us."[56]

What could be said? Solzhenitsyn, they claimed, had uninformed impressions of American democracy and was ill-equipped to propose solutions. He misunderstood the nature of American institutions, such as the courts and the press, and his plea for asceticism smacked of the seventeenth century. Even someone who was sympathetic to his courageous warning about totalitarianism, such as Sydney Hook, the ex-Marxist political philosopher, discounted his religious and prophetic stance with a touch of condescension and ridicule, as, unsurprisingly, did the historian Arthur Schlesinger Jr.[57]

The one wholeheartedly enthusiastic response came from the Catholic theologian and publicist Michael Novak, who called the Harvard address "the most important religious document of our time." Novak compared the idea of a "wrong turn" in the Renaissance and the Enlightenment to the ideas of T. S. Eliot, Paul Tillich, Christopher Dawson, Emil Fackenheim, Richard Weaver, Eric Voegelin, Leo Strauss, and others, both Jewish and Christian. Novak applauded Solzhenitsyn's critique of moral and religious individualism—"You have your moral convictions, I have mine, who can tell who is right?"—which leads to the impasse in which "millions . . . have declared themselves incompetent to make value judgments." And this declaration leads to an attrition of *all* values except instinctual ones. (Simone Weil had observed the same poverty of public values in France a generation earlier.)[58]

Solzhenitsyn does not clearly indicate what religious correction he would make in American society—whether he thinks that high values are compatible with religious pluralism or whether he seeks ecumenically to find common values in several faiths, such as those represented by Michael Novak, Reinhold Niebuhr, and Will Herberg.[59]

## vi

In 1983, Solzhenitsyn was awarded the Templeton Prize for "Progress in Religion" by the British royal family at Buckingham Palace. He expressed his gratitude for the recognition given for the first time to a member of the Orthodox Church. (Mother Teresa and Billy Graham had received the prize before.) He also reminisced about the church he attended in Kislovodsk, which had been looted before worshipers' eyes, and about being harassed by playmates for frequently attending church. After the First World War, he said, the Orthodox Church was torn apart and "subjected to persecution even fiercer and more extensive than that of early Christian times."[60]

Still, he was more concerned about his adopted West than about Russia. He looked to the West to fight a war against communism, not with atom bombs but with repentance and conversion. Five years after his Harvard appearance, however, he was even more pessimistic about the West—the problem now, he indicated, was the slide not just into materialism but into atheism. "Atheist teachers in the West," he said, "are bringing up a younger generation in a spirit of hatred of their own society." He cited Dostoyevsky's view that the world would be saved only after a "visitation by the demon of evil." The revolution, he said, began with atheism, which is the "pivot" of

Marxism, rather than with economics. Along the way, Solzhenitsyn never hesitated to make invidious comparisons with the West to puncture its balloon of complacency. Japan and Free China "have preserved their moral sensibility to a greater degree than the West has, and have been less affected by the destructive spirit of secularism," he stated. Even the World Council of Churches, he said bluntly, was more concerned with food distribution than religious persecution in Russia. And Billy Graham himself was silent about or unaware of it.[61]

There was hope nonetheless. Orthodoxy had survived the Revolution, "for no matter how formidably Communism bristles with tanks and rockets, no matter what successes it attains in seizing the planet, it is doomed never to vanquish Christianity." In this speech at Buckingham Palace, the term "atheism" is used a dozen times, and "secularism" four times, the latter transmogrifying into the former.[62]

Solzhenitsyn's last major address before returning to Moscow was delivered to the International Academy of Philosophy at Liechtenstein on September 14, 1993, two years after the fall of communism in Russsia. He refers to the communist collapse as "the disintegration of the world's most senseless, recklessly wasteful economy." Acknowledging Reagan's "new spiraling and ultimately unbearable arms race," he observed that communism had broken down with incredible speed in a dozen countries simultaneously and that the nuclear menace was vanishing.[63]

Why, then, was this not an occasion to rejoice? Because the "problems" that Solzhenitsyn had been outlining for twenty years in the West had not gone away. The theater of war was still the West: materialism, which can never satisfy; the illusion of progress; the gigantesque consumption of the world's resources by the sole remaining megapower; the horrible poverty of most of the rest of the world; overpopulation (the solution to which he does not give); and the "brazen use of social advantage and the inordinate power of money." Dipping into modern philosophy once more, he blamed Locke and Bentham respectively for the separation of politics from moral criteria and for hedonism. If these ills were not attended to, "then the vast red whirlwind may repeat itself in its entirety." Mankind would have no future of which to speak, he warned.

But, finally, he did recognize the unique and positive values and achievements in America, which guaranteed "stability of civic life under the rule of law—a hard-won stability which grants independence and space to every private citizen." Solzhenitsyn clearly valued the freedoms afforded to him by his adoptive country. He only feared that their abuse and exaggeration would lead to their demise.[64]

Unique to Solzhenitsyn's address to the West was his personal testimony to the relatively hidden and horrendous sides of communism: the suppression of all freedom, speech, and religion. Solzhenitsyn had lived through its worst phase and could all but show the wounds in his hands and his side. He had suffered, but he had overcome. He had outsmarted the Soviets. He had created a network of "invisible allies" who were willing to face imprisonment and physical injury to get him published abroad. This accomplishment gave him the leverage of widespread fame and the certainty that his mysterious disappearance would not go unnoticed.

The Solzhenitsyn phenomenon is an incredible life-story that stood testimony to a spiritual reawakening culminating in 1989 and 1991. For Solzhenitsyn, secularism—the removal of God from public life—lay at the core of Soviet and world communism. Property and nuclear weapons were secondary.

His *Warning to the West* is that this disease is already in an advanced stage and could overtake it—soon. The collapse of the Soviet Union in 1991 did not eliminate this danger, but only the geography of aggression that was more moral than military. The problem was the ideology that Dostoyevsky called the "fire in the minds of men."[65]

# Chapter Fourteen
# Conclusion

Why should secularism be considered a problem when it was thought to have been a solution? Christianity, as we know, substituted the "this-worldly" values of the ancient city for transcendental ones. But this did not necessarily result in the rejection of much that was classical Rome's. So Augustine urged the coexistence (*permixtio*) of the two cities, the city of God and the terrestrial city. The *saeculum,* and its classical culture, he thought, were not evil in themselves, but rather the proving ground of the heavenly city.

A central problem of the Middle Ages was the papal claim of *plenitudo potestatis,* or the superiority of the pope over temporal rulers on account of the superior nature of the papal mission: eternity over time. But beginning in the fourteenth century, lay authors like Dante began to deny this superiority by idealizing the German emperor and classical Rome. Two centuries later, Machiavelli depicted a prince freed of Christian moral constraints. Both he and Dante would secularize Christendom by limiting the authority of its pontiff and revalorizing antiquity.

The consequence of late medieval secularism became more evident in the thought of the Protestant reformers' total rejection of the papacy's suzerainty over Christendom. Luther's *sola scriptura* opened the door to private interpretation and sectarianism—a major consequence of Protestantism, which fueled the English civil wars. Locke concluded that no public orthodoxy was possible because no one religion could claim to be the true one. Locke, in our view, is the father of modern secularism.

On the Continent, the attempt to preserve orthodoxy was more often than not secular. Jean Bodin (1530–1596), at the end of the religious wars between Protestants and Catholics, argued that there could be no appeal beyond a sovereign (king). Louis XIV forced religious uniformity on his subjects while simultaneously declaring his kingdom independent from Rome in all but doctrinal matters.

Rousseau accents Locke's privatization of the Christian conscience, but he thinks that it must be combined with a (pagan) civic state religion, modeled on that of republican Rome. The French Revolution sponsored an outdoor religion of civic festivals, responding partly to the new appreciation of nature and to the old love of the ancient republic. Kant exceeded Rousseau's secularism by first denying the rational foundations of the truths of religion. Both Rousseau and Kant undercut the ontological claims of religion, which were forfeited on behalf of the interests of the secular state and the individual conscience. Private and public *personae* were thus severely bifurcated.

Hegel sought to reconcile this split among Christianity, Enlightenment, and Freedom. The Middle Ages, he felt, had alienated man, resulting in the "unhappy consciousness." Reason working through the dialectic of history, and struggling with every material, spiritual, and historical obstacle, could objectify freedom in the state and attain pure self-consciousness, or Absolute Spirit. Hegel saw this process as contained within time, that is, immanent rather than transcendent. Humanity is elevated to divinity largely by denying the distinction between the two and by discovering that evil always works for the good—that is, by not being really evil. With his follower Ludwig Feuerbach, the process is further immanentized: man creates God in his own image and alienates himself in the process, or he repossesses God by making him part of himself.

Marx, in his turn, believed that the atheistic, communist state was the agent of emancipation not only from capitalism but also from religious denominationalism, both Jewish and Christian. After the Second World War, Marxist-communist states, which were built on Marx's thought, claimed to be "popular" dictatorships of the people over themselves, or "popular republics." Converting God's sovereignty into human self-dominion, man then surrendered himself to the state. Or as Rousseau put it, he allowed himself to be "forced to become free."

Battling severe emotional and physical handicaps, Dostoyevsky became secularism's most perspicacious observer and prophet, for he had a keen vision of where the European secular movement was going. The logical consequences of its denial of God ended up in Nechaev's nihilism. Once immortality is denied, everything is permissible and the immediate result is the total expendability of human life, including one's own.

The totalitarian state can be seen as an attempt (at least on the part of its elites) to fill the vacuum left by Christianity's decline. Ever since the early nineteenth century, many socialist philosophers such as Henri de Saint-Simon had thought of a "new Christianity," which, unlike the old, would eradicate poverty and bring heaven to earth. It espoused a purely secular faith, in which everyone was Orthodox. The state took the place of the church and promised happiness, here, rather than in eternity. The church was not necessary because original sin simply did not exist. Man is perfectible, as Condorcet had said in 1794.

The twentieth century brought millions back to the catacombs, such as the early Christians—that is, to the Gulag that Solzhenitsyn described in one of the most poignant testimonies of the contemporary world. The Gulag was not just a Soviet prison camp, because Solzhenitsyn claimed that it was borrowed by Hitler to erect Germany's concentration camps. Rather, the Gulag was a gigantesque network of forced labor camps, or "archipelago," extended through hundreds and thousands of square kilometers in Siberia, where, under atrocious conditions, millions were forced to work, live, and die for having committed often imaginary "crimes" against Stalinism. Solzhenitsyn believed that the Soviet experience warranted a warning to the West. How is this conceivable? It seems as improbable as was the thought of the Bolshevik Revolution in 1913. Totalitarianism, however, can grip men's minds when, having come to believe nothing, they are willing to believe anything, to paraphrase G. K. Chesterton. What Raymond Aron called "the opium of the intellectuals" can nourish a belief as convincing and encompassing as the one that has been lost. In this light, it is quite possible to see that totalitarianism was as complete a belief system as monotheism had been in the past. Solzhenitsyn called it an "ideology," which we can define here as an unreflective mental habit that locks everything mechanically in place and justifies a grid—of secret police, torturers, guards, stool pigeons, psychiatrists, and others who are not fully responsible—carrying out orders automatically in the Gulag civilization.

Taking up the cause of the industrial proletariat was not left solely to Marx and the totalitarian regimes that invoked his authority. There were many others, among them, Simone Weil, whose spirited and difficult life of nonviolent activism embraced the cause of both social justice in factory work and religious transcendence in the Catholic Church. Mother Teresa and Saint Josemaría Escrivá both worked among the poor and attracted the rich. Work done with love rather than invidious class hatred is what sanctifies. Social causes were popular with thousands in the United States and among left-wing liberation theologians in Latin America. The danger, however, was that Christianity could be used exclusively as a social program.

While the Marxist totalitarianism of the Cold War seems remote to a post-9/11 culture, the capitalism that Marx attacked in *Das Capital* is being attacked today by Muslims who target its American outposts of wealth, power, and weapons. Many Muslims, then, view secularism as an inimical Western, and even Christian, phenomenon that needs to be extirpated by Islam.

The modern Western divorce between religion and the church, religion and the state, and religion and happiness reached acute dimensions in the twentieth century. John Paul II has called it the century of death in his best-selling antidote *Crossing the Threshold of Hope* (1994). A recomposition of life in which the religious and the secular coexist and interpenetrate peacefully is a solution worth trying in the future. It is one being sought by Protestant evangelists and American Catholics in the United States, but so far rejected by the European Union.

The experience of Western civilization has taught us the important distinction between the religious and the secular (indeed, this distinction is one of its defining characteristics). One benefit derived from it is the lesson of toleration of different religions, which Locke had advocated in 1689. The flip side is the elimination of all tolerant religions from the public square until they are replaced by nontolerant religions or ideologies that are again on the rise. The problem still exists, however, as Locke saw it (the Le Pen movement in France and the worldwide Islamist movement are two examples).

What can the present analysis bring to these conflicting eddies of secularism and its critics? Secularism can be elucidated by the Aristotelian concept of virtue: the mean between two extremes. Augustine's disengagement from the ancient city represents one pole in the history of secularism, although even Augustine never fully forsook the *saeculum*. Aquinas, for his part, while endorsing the superiority of the spiritual over the temporal, had enough confidence in human nature and believed that religion need not be wholly dependent on grace and faith. St. Thomas represented the medium point in the secular spectrum. Marx represented the opposite extreme, because he did not acknowledge the right of the Christian or the Jewish community to exist, but would dissolve them into one commonality, with no religion, property, or family to divide them.

From Aquinas in the High Middle Ages down to the present, the path is not straight, but retrospectively it certainly points in the direction of contemporary secularism. By retracing our steps down this path, we can reach a better understanding of the way we are and increase our options about the way we could be.

Certainly, in the twentieth century, there has been a return of many to Orthodoxy and a move away from biblical historicism and theological modernism, often, as in the case of Karl Barth, marked by the catastrophe of world war. Rational, liberal, patronizing, and bourgeois biblicism gave way to the far more penetrating and disarming apologetics of Martin Buber, Karl Barth, and C. S. Lewis, who quit the historical "de-mythologizing" and assumed a direct (existential) relationship to God. Meanwhile, philosophers such as Jacques Maritain, Etienne Gilson, and Joseph Pieper proved prolifically that Aristotelian Thomism was still a valid alternative to contemporary irrationalisms.[1]

One practical option traced by a recently canonized saint is particularly relevant to secularity, if not secularism. Josemaría Escrivá de Balaguer (1902–1975) founded Opus Dei (the Work of God) in 1928 and has drawn to his way hundreds of thousands of lay men and women on five continents who seek to sanctify their ordinary work. Escrivá explained it to the *New York Times* in 1966: "In God's service there are no second-class jobs; all of them are important. To love and serve God, there is no need to do anything strange or extraordinary. Christ bids all men without exception to be perfect as his heavenly Father is perfect."[2] Opus Dei has proved to be one of the most remarkable realizations of Vatican II's call for "universal holiness." Stressing the materialization of the spiritual, it may well present the synthesis between the sacred and the secular that the modern world lacks. The personalist dimension of its message is very much like Mother Theresa's, except that it is usually carried out in the ordinary (secular) circumstances of the workplace, rather than in the extraordinary circumstances of destitution. Perhaps we are coming full circle—to the time of the early Christians, of St. Augustine, who stressed the fusion of the two cities (*permixtio*) and the worthiness of ordinary Christians who performed manifold secular occupations in the Roman Empire. The Benedictine "orare est laborare" was an ideal of centuries of monasticism. It is necessary to call the modern way of life back to its foundations in a modern way, incorporating into it the high ideals, skills, practices, institutions, and wonders of recent experience. Over the past two millennia, Western man has had great difficulty integrating faith and reason, the sacred and the secular. Let us not hear again the reproach Simone de Beauvoir made of the victors of the Second World War in *The Mandarins:* "They gave us back our lives, without giving us a reason to live."

# Notes

## Chapter 1

1. Numa Denis Fustel de Coulanges, *The Ancient City,* foreword A. Momigliano and S. C. Humphreys (Baltimore: Johns Hopkins University Press, 1980). A short list of some of the most important titles on Western secularism would include the following: Hans Blumenberg, *The Legitimacy of the Modern Age,* trans. Robert M. Wallace (Cambridge, MA: MIT Press, 1983; orig. publ. 1966); the special issue of *Daedalus* (Summer 2003) dedicated to secularism; Keith Thomas, *Religion and the Decline of Magic* (New York: Scribner's, 1971); James E. Cummins, *Religion, Secularization and Political Thought: Thomas Hobbes to J. S. Mill* (London and New York: Routledge, 1989); Lewis W. Beck, *Six Secular Philosophers* (London: Thoemmes Press, 1997; orig. publ. 1960); W. Warren Wagar, ed., *The Secular Mind: The Transformation of Faith in Modern Europe: Essays Presented to Franklin L. Baumer* (New York: Homes and Meier, 1982); Alasdair MacIntyre, *After Virtue: A Study in Moral Theory* (Notre Dame, IN: Notre Dame University Press, 1981); Owen Chadwick, *The Secularization of the European Mind in the Nineteenth Century* (Cambridge: Cambridge University Press, 1975); Stephen Carter, *The Culture of Disbelief* (New York: Basic Books, 1993); Peter Berger, *Facing Up to Modernity: Excursions in Society, Politics, and Religion* (New York: Basic Books, 1977); *A Far Glory: The Quest for Faith in the Age of Credulity* (New York: Free Press, 1992); "The G-Word and the A-List: In a Social Setting, There's One Subject Washington Avoids Religiously: God," *Washington Post,* July 12, 1999, C1. The present work differs from most of the above outstanding studies, in that it searches for the origins of the secular problem in Christian antiquity and the Middle Ages. Additionally it considers many of the modern authors from the aspect of secularism for the first time, and finally it faults the contemporary secular arrangement as myopic.

## Chapter 2

1. Aegidio Forcellini et al., *Lexicon totius latinitatis* (Padua, 1940); Augustine, *De doctrina Christiana,* ed. W. H. Arnold and Pamela Bright (Notre Dame, IN: University of Notre Dame Press, 1995); Augustine, *Against the Academicians: The Teacher,* ed. and trans. Peter King (Indianapolis, IN: Hackett, 1995), 189–190; Augustine, *Christian Instruction [De doctrina Christiana],* trans. John J. Gavigan, O.S.A. (New York: Cima, 1947); Augustine, *Concerning the City of God against*

*the Pagans*, trans. Henry Bettenson, intro. David Knowles (Harmondsworth, UK: Penguin Books, 1972); Augustine, *Works*, 15 vols., ed. Rev. Marcus Dods, trans. J. G. Cunningham (Edinburgh: T. & T. Clark, 1872–1934), vol. 13 (1875); Henri Xavier Arquillière, *L'Augustinisme politique; essai sur la formation des théories politiques du Moyen Age*, 2nd ed. (Paris: J. Vrin, 1955); Peter Brown, *Augustine of Hippo: A Biography* (London: Faber and Faber, 1969; orig. publ. 1967); Brown, *The Body and Society: Men, Women and Sexual Renunciation in Early Christianity* (New York: Columbia University Press, 1988); Brown, *Power and Persuasion in Late Antiquity: Towards a Christian Empire* (Madison: University of Wisconsin Press, 1992); *Catalogus verborum quae in operibus Sancti Augustini inveniuntur, Thesaurus Linguae Augustinianae* (Eindhoven, the Netherlands: Augustijnendreef, 1976–1993), vols. 6 and 7; Allan D. Fitzgerald, O.S.A., ed., *Augustine through the Ages: An Encyclopedia* (Grand Rapids, MI: W. B. Eerdmans, 1999); Numa Denis Fustel de Coulanges, *The Ancient City: A Study on the Religion, Laws and Administration of Greece and Rome*, foreword A. Momigliano and S. C. Humphreys (Baltimore: Johns Hopkins University Press, 1980); C. N. Cochrane, *Christianity and Classical Culture* (New York: Oxford University Press, 1957), 129; Etienne Gilson, *La philosophie au Moyen Age, des origines patristiques à la fin du XIVe siècle* (Paris: Payot, 1947); Gilson, *Reason and Revelation in the Middle Ages* (New York: Scribner's, 1938); R. A. Markus, *Saeculum: History and Society in the Theology of St. Augustine* (Cambridge: Cambridge University Press, 1970); Markus, "The Sacred and the Secular: From Augustine to Gregory the Great," *Journal of Theological Studies* n.s. 36 (1985): 84–96; R. I. Moore, *The First European Revolution* (London: Blackwell, 2000); Arpad Péter Orban, *Les Dénominations du monde chez les premiers auteurs chrétiens* (Nijmegen: Dekker and Van de Boegt, 1970); Robert Estienne, ed., *Lexicographorum principis Thesaurus linguae latinae*, 4 vols. (Basil: E. & J. R. Thurnisiorum frat., 1740–1743); Gary Wills, *Saint Augustine's Childhood: Confessions*, bk. 1 (London: Continuum, 2001); Langdon Gilkey, "Ordering the Soul: Augustine's Manifold Legacy," *Christian Century*, April 27, 1988, p. 427; Brown, *Augustine*, 20, 31. Fustel de Coulanges's *Ancient City* describes this all-encompassing, self-sufficient community as having a religion that was particular to it and that dominated every aspect of public and private life.

2. Augustine, *Confessions*, trans. F. J. Sheed, 2nd ed. (Indianapolis, IN: Hackett, 2006), VIII, xii; II, iv. See also Wills's delightful *Saint Augustine's Childhood*, bk. 1.
3. Fitzgerald, 39–41.
4. On the question of the body in Augustine's works, see Brown, *Body and Society*, chap. 19.
5. Augustine, *Four Anti-Pelagian Writings*, trans. John A. Mourant and William J. Collinge (Washington, DC: Catholic University Press, 1992), 25–105, 218–270.
6. Fitzgerald, 34–39.
7. *Conf.* I, 1.
8. Forcellini et al., s.vv. *saecularis, saeculum*, 176, 196, 225, 187, 189, 191, 234, and passim; John 15:18–20; 17:9–19; Estienne, 4:143. Most clerical uses of the word cited here are neutral, meaning "a hundred years." A few mean "the age" and are occasionally derogatory. Orban, 234, 198, 211, 221, 167, 169,

Notes / 233

191, 176, 196, 198, 225; Gilson, *Reason and Revelation*, 8–10, 12; and Gilson, *La philosophie au Moyen Age*, 98.
9. Cornelius Mayer, ed., *Augustinus-Lexicon*, vol. 1; Brown, *Catalogus verborum*, s.vv. *saecularis, saeculum.*
10. Brown, *Augustine*, 291–295; Augustine, *City of God*, I, xiv; XIV, i, iv.
11. Brown, *Power and Persuasion.*
12. Augustine, *City of God*, XIV, xxviii; XIX passim, xxxii; ibid., *Confessions*, I, i.
13. Augustine, *City of God*, XIV, iii, ix.
14. Ibid., I, 35.
15. Augustine, *Epistle*, CXXXVIII, in *Works*, 13:206.
16. Romans 13:1–7; 1 Peter 13–18.
17. Augustine, *Works*, V, xvii (my emphasis).
18. The Dante scholar John Freccero observed in a note to me of May 1963 that St. Augustine's political theory was "subversive." See Arquillière, passim; Cochrane, 129; Markus, *Saeculum*, 54–55. Markus argues that the *saeculum* is neutral ground between the city of God and the terrestrial city and that it is neither Christian nor entirely corrupt. Thus it is not incorporated into a Christian order. Indeed the attempt to do so after A.D. 476 spelled disaster for the West. Christianity, he felt, should not be "established." My interpretation is that the *saeculum* is identical with the terrestrial city, i.e., a fallen world in which Christians could comingle but not belong entirely.
19. Augustine, *City of God*, XIX, 4, 5, and passim. In XV, 26, he refers to life *in saeculo maligno*; cf. XIX, 27.
20. Augustine, *Against the Academicians*, 189–190; *Christian Instruction*, 95, 111, 74. See also the collection of essays by C. Scharblin and F. van Fleteren in Duane W. H. Arnold and Pamela Bright, eds., *De doctrina Christiana* (Notre Dame, IN: University of Notre Dame Press, 1995), 14–24, 47–67.
21. Markus, "The Sacred and the Secular," 84–96. Brown's *Power and Persuasion* documents the interaction of Christian bishops with imperial authorities on such matters as religious toleration and poor relief. The bishop generally protected those who were poor and not citizens.
22. Moore, 122.
23. Markus, *Saeculum*, passim.
24. Ernst Troeltsch, *The Social Teachings of the Christian Churches*, trans. Olive Wyon, 2 vols. (Louisville, KY: Westminster John Knox Press), passim; Markus, *Saeculum*, 53–55, 126, 151–152, 181–184; Cochrane, 129; Brown, *Power and Persuasion*, passim; Ramsay MacMullen, *Christianity and Paganism in the Fourth to Eighth Centuries* (New Haven, CT: Yale University Press, 1997), 155 and passim.
25. Markus, *Saeculum*, 133–143, chaps. 5 and 6 passim.

## Chapter 3

1. The philosophy department of George Washington University, for example, jumps from Phil. 111, "History of Ancient Philosophy," to Phil. 112, "History of Modern Philosophy." The main sources of this chapter are as follows: Thomas Aquinas, *Summa theologica*, trans. Fathers of the Dominican Province,

5 vols. (Allen, TX: Christian Classics, 1981; orig. publ. 1911)—henceforth cited as *S.T.*, followed by the number of the question and then the article of the question; *Expositio super librum Boethii de Trinitate*, ed. Bruno Decker (Leiden: E. J. Brill, 1965); *Summa contra Gentiles*, trans. and ed. C. Pegis et al., 4-vol. ed. (South Bend, IN: Notre Dame University Press, 1975); *Treatise on Kingship to the King of Cyprus*, trans. Gerald B. Phalen, ed. I. T. Eschmann (Toronto: Pontifical Institute of Medieval Studies, 1949).

2. Marcia Colish, *Medieval Foundations of the Western Intellectual Tradition* (New Haven: Yale University Press, 1997), 167–171, 175–222, 265–273, 319, 295; R. L. Moore, *The First European Revolution, c. 970–1215* (London: Blackwell, 2000), 149–159. The first revisionist to champion the twelfth century was Charles Homer Haskins, *The Renaissance of the Twelfth Century* (Cambridge: Harvard University Press, 1976; orig. publ. 1927).

3. John P. Wippel, "The Condemnation of 1270 and 1277," *Journal of Medieval and Renaissance Studies* 7 (1977): 169–201; Jacques Maritain, *The Angelic Doctor: The Life and Thought of Saint Thomas Aquinas*, trans. J. Scanlan (New York: Dial Press, 1931).

4. Maritain, 30–35; Gilson, *The Philosophy of Thomas Aquinas* (Salem, NH: Ayer Co., 1985; orig. publ. 1924), chap. 1; Martin Grabmann, *Thomas Aquinas, His Personality and Thought* (New York: Longmans, Green & Co., 1928), 2–15. Thomas was in Naples between 1260/61 and 1272/73, and in Paris between 1253–1259 and 1268–1272.

5. Thomas Gilby, *The Political Thought of Thomas Aquinas* (Chicago: Chicago University Press, 1973; orig. publ. 1958), 101.

6. Maritain, 41.

7. Gilson, *Reason and Revelation in the Middle Ages* (New York: Scribner's, 1938), chaps. 1, 2, 24. Cf. Augustine's formula: "Therefore seek not to understand that thou mayest believe, but believe that thou mayest understand," ibid., 19.

8. *De Caelo et de Mundo* cited in E. J. Dijksterhuis, *The Mechanization of the World Picture* (Princeton: Princeton University Press, 1964), 133.

9. Quentin Skinner, *The Foundations of Modern Political Thought*, vol. 1, *The Renaissance* (Cambridge: Cambridge University Press, 1996; orig. publ. 1978), 105–110.

10. Fernand Van Steenberghen, *Thomas Aquinas and Radical Aristotelianism* (Washington, DC: Catholic University Press, 1980), 1–27; Aquinas, *S.T.* I, 44, 45.2, 46.2.

11. Aquinas, *Expositio super librum*, 94.

12. Ibid.; *S.T.* I, 46.

13. Jan Miel, *Pascal and Theology* (Baltimore: Johns Hopkins University Press, 1969)—an excellent discussion of the Augustinian tradition of grace and free will. Luis Molina (1535–1600) was a Spanish Jesuit who attempted to reconcile free will with divine grace.

14. J. Deferrari, *A Latin Dictionary of St. Thomas Aquinas Based on the "Summa theologica" and Select Passages in Other Works* (St. Paul Editions, 1960), 930–937; Robert Busa, *Index Thomisticus: Sancti Thomae Aquinatis operum omnium indices et concordantiae* (Rome, 1975), s.v. *saeculum*. On Molinism see R. R. Palmer, *Catholics and Unbelievers in Eighteenth Century France* (Princeton: Princeton University Press, 1939), 31ff.

15. *Summa contra Gentiles*, I, 13.
16. *Summa contra Gentiles*, passim; Gilson, "What is Christian Philosophy?" in *A Gilson Reader* (Garden City, NY: Image Books, 1957), 183–185; Gilson, *Reason and Revelation*, passim; *S.T.* passim.
17. Gilson, *Philosophy of Thomas Aquinas*; chap. 4 is devoted to the five proofs of the existence of God.
18. "Muddling along" was Freud's phrase. "Absurdity" was Sartre's and Camus'.
19. Gilby, 101; Genesis 1:31.
20. *S.T.* II, II.188.
21. Geoffrey Barraclough, *The Origins of Modern Germany* (New York: Capricorn Books, 1963; orig. publ. 1948), 52, 61; *The Correspondence of Pope Gregory VII: Selected Letters from the Registrum*, trans. and ed. Ephraim Emerson (New York: W. W. Norton, 1960; orig. publ. 1932), 48, 124.
22. Jose Sanchez, *Anticlericalism: A Brief Study* (Notre Dame, IN: University of Notre Dame Press, 1972), chap. 3.
23. John Finnis, *Aquinas: Moral, Political and Legal Theory* (Oxford: Oxford University Press,1998), 188–196; *S.T.* II, 77.4; 187.2; 184.8, 6.
24. Luke 10:41–42.
25. Aquinas, *Treatise on Kingship*, passim.
26. Jean de Joinville, "The Life of St. Louis," in Chronicles of the Crusades, ed. Joinville and Villehardouin, trans. M. R. B. Shaw (Harmondsworth, UK: Penguin Books, 1963), chaps. 12, 18, 20.
27. *S.T.* II, IIae, 63; 187; Fritz Kern, *Kingship and Law in the Middle Ages*, trans. S. B. Chrimes (New York: Harper, 1970; orig. publ. 1956), 39.
28. Brian Tierney, *The Crisis of Church and State, 1050–1300* (Englewood Cliffs, NJ: Prentice Hall, 1964), 49–50, for Gregory VII's *Dictatus papae*. The deposition of Frederick II (who defended himself by natural law) by Innocent IV (ca. 1246) is described in terms of "plenitude of power," ibid., 147. On decretalists on same theme, ibid., 153–157. Walter Ullmann, *Papal Government in the Middle Ages: A Study in the Ideological Relation of Clerical to Lay Power* (London: Methuen, 1955).
29. *S.T.* II, IIae, 60.6; 104.6.
30. Aquinas, *Treatise on Kingship*, I, i; II, vi; II, iii.
31. Ibid., I, x, xi.
32. Ibid., II, iii.; Bernard Plongeron, *Théologie politique au siècle des lumières (1770–1820)* (Geneva: Droz, 1973), chap. 2.
33. *S.T.* I, IIae, 27.1 (my emphasis), 34.4; Augustine, *Confessions*, I, i (my emphasis).
34. Ibid., I, IIae, 94.1, 19.5,6, on erring conscience.
35. Ibid., II, IIae, 11, 39.
36. Ibid., IIae, 19.5,6. On natural law and the "old law," see I, IIae, 94, 98, and "Treatise of Law," ibid., 90–100. For a good introduction to the history of natural law, see Richard Tuck, *Natural Rights Theories: Their Origin and Development* (Cambridge: Cambridge University Press, 1979). Aquinas tended to speak of *ius* as objective law rather than subjective right as in later usage, but he did formulate a concept of property right as individual; see Brian Tierney, *The Idea of Natural Rights* (Atlanta: Scholars Press, 1997), 146, 256–265. The whole concept of "natural rights," especially as it developed in the seventeenth century, is secular insofar

236 / NOTES

as rights are derived from philosophy rather than theology, and with reference to the layperson and the state rather than to the clergy or the church.
37. *S.T.* I, IIae, 109.2,5,7.b.

## Chapter 4

1. In the *Summa theologica*, pt. I–II, q. 25, art. 2, Aquinas writes, "We name a thing as we understand it, for *words are signs of thoughts* as the Philosopher [Aristotle] states . . . Now in most cases we know a cause by its effect. But the effect of love, when the beloved object is possessed, is pleasure; when it is not possessed, it is desire or concupiscence." Cf. Dante, *The Divine Comedy*, trans. and ed. Charles S. Singleton, 6-vol. ed. (Princeton University Press, 1970–1975), vol. 1, pt. 1: *The Inferno*, I, 49–64. The main sources used in this chapter are *The Divine Commedy*; *Vita Nuova*, new trans. Mark Musa (Bloomington: Indiana University Press, 1973; orig. publ. 1957). Citations to the *Commedia* are to Singleton's bilingual text, by cantiche (e.g., *Inferno*, hereafter abbreviated *Inf.*; *Purgatorio*, *Purg.*; and *Paradiso*, *Par.*). Numbers thereafter refer to canto and line. Each cantiche of this edition is accompanied by a separate volume of commentary by Charles Singleton, designated as pt. 2. Singleton amply documents Dante's Thomism throughout the *Commedia*. For the opposite view, see T. S. Eliot, *Dante* (London: Faber and Faber, 1929), 44; cf. Daniel J. Ransom, *Dante Studies* 95 (1977): 81–99. Thomas C. Maresca, "Dante's Vergil: An Antecedent and St. Augustine," *Neophilologus* 65 (1981): 548–555; *A Translation of the Latin Works of Dante Alighieri* (New York: Greenwood Press, 1969; orig. publ. 1904); *Monarchy*, trans. and ed. P. Shaw (Cambridge: Cambridge University Press, 1996); *Dante's Rime*, trans. and ed. Patrick S. Diehl (Princeton: Princeton University Press, 1979).
2. *Vita Nuova*, xxi, ii, xxxi; *Rime*, no. 13. The quotation is attributed by Dante to Homer, whom Dante could not have read directly: G. Boccaccio, "Life of Dante," in *The Earliest Lives of Dante*, trans. James Robinson Smith (New York: Russell and Russell, 1968, orig. publ. 1901), 19; William Anderson, *Dante the Maker* (London: Routledge / Kegan Paul, 1980), 68–75. The term "dart" can be found in *Rime*, no. 13.
3. *Vita Nuova*, xxxi.
4. *De Vulgare Eloquentia*, trans. and ed. Steven Botterill (Cambridge: Cambridge University Press, 1996); Singleton, *Essay on the Vita Nuova* (Baltimore: Johns Hopkins University Press, 2000; orig. publ. 1949), chap. 3.
5. Eric Auerbach, *Dante: Poet of the Secular World*, trans. Ralph Mannheim (Chicago: Chicago University Press, 1961), 921; Singleton, *Essay*, chap. 3.
6. *Itinerarium mentis ad Deum*, quoted in Singleton, *Essay*, 39.
7. Hans Baron, *The Crisis of the Early Italian Renaissance* (Princeton: Princeton University Press: 1999).
8. Anderson, chaps. 10 and 13.
9. Anderson, chap. 10; Joan M. Ferrante, *The Political Vision of the Divine Comedy* (Princeton: Princeton University Press), 69, 158, 195, 286, 351.
10. *Inf.* III, 59.
11. J. N. D. Kelly, *The Oxford Dictionary of the Popes* (Oxford: Oxford University Press, 1986), 208–212.

12. *Rime*, no. 88a; Anderson, chap. 13 and passim.
13. *Inf.* VI, 74–78; Anderson, chap. 10 and passim.
14. Dante, "Epistles," in *Translation of Latin Works*, 300.
15. Letter to a Florentine Friend, June 2, 1316; "Epistles," 341.
16. Lorenzo Valla, *The Treatise on the Donation of Constantine* (New Haven: Yale University Press, 1922). The donation was supposed to have confirmed papal ownership of lands given to it in the later Roman Empire. By the late fourteenth century it was thought to include today's Latium, Umbria, the Marches, and Emilia-Romagna.
17. W. H. Reade, "Political Theory to c. 1300," in *Cambridge Medieval History*. ed. J. R. Tanner, C. W. Previté-Orton, and Z. N. Brooke (New York: Macmillan, 1911–1936), 6:602–638; "Unam Sanctam," in *Classics of Western Thought*, vol. 2, ed. F. Thompson (San Diego: Harcourt Brace, Jovanovich, 1988; orig. publ. 1964), 89–90. Philip IV and Boniface appear in *Inf.* XIX; Boniface again in XXVII, 70.
18. Dante, *Monarchy*, trans. and ed. P. Shaw (Cambridge: Cambridge University Press, 1996), I, ii, v. All citations to *Monarchy* are to this edition by book and section.
19. Ibid., II, ii, iii, x.
20. Ibid., II, iii–v.
21. Ibid., II, v.
22. Ibid., III, iii, iv, viii–x, xiii. Charles Till Davis, *Dante and the Idea of Rome* (Oxford: Clarendon Press, 1957), emphasizes the essentially negative idea Augustine had of pagan Rome, from which the best that could be expected is that it let Christians alone. In contrast to that negative view, Dante sees the instrumentality of the Roman Empire in preparing the way for the coming of Christ and the spread of Christianity. For Augustine, Vergil, the poet, whose Aeneas founded Rome, was not representative of reason. But for Dante, Vergil's fourth eclogue announced the birth of Christ. Rome is providential for Dante. However, it is true that Augustine in Ep. 138 (see above) stresses the good citizenship of Roman Christians and therefore the compatibility of Christianity with the Roman Empire. But "Augustine tried to demolish the myth of a universal and eternal Rome" (62).

While some authorities in the Middle Ages regarded Rome as a sacerdotal *and* regal city, the home of popes and emperors, Dante saw it primarily as the seat of the Roman emperor, providential in preparing the matrix of Christianity, and universalist as representing the entire human race, rather than particularist like the Italian city-states. Davis writes, "Dante's Monarch is at the same time Vergilian and Augustinian, Roman and Christian and his Empire is both a *remedium peccati* (remedy of sin) and an instrument for realizing the highest potentialities of the human race" (141). "Dante in his ardour did not realize the impossibility of restoring Italy to her ancient place as the jurisdictional heart of the Empire, and second, that even in those cities where Henry was able to impose his *regimen*, it did not bring the unity and peace which the Florentine exile expected," 176. For another more recent view of this matter, see Giuseppe Mazzotta, *Dante, The Poet of the Desert: History and Allegory in the "Divine Comedy"* (Princeton: Princeton University Press, 1979), chap. 4; Luke 22:38.

23. *Monarchy*, III, xii. Aegidius Romanus [Giles of Rome], a follower of St. Thomas Aquinas, in his *De Ecclesiastica Potestate*, ed. R. W. Dyson (Suffolk, UK: Bydell Press, 1986), written shortly after the *Unam Sanctam* of Boniface VIII, states, "It is fitting for the Church to possess the material sword, not as user, but as commander." (I, ix). Explaining this further, he writes, "If neither our end nor our hope of blessedness is to be placed in temporal things, but in spiritual, then it must be that temporal goods are not good except insofar as they are ordered towards spiritual ends" (II, iv, 4). So the pope, who represents the spiritual power, "has power over souls, over bodies and over the temporal goods which [persons] possess." As a cautionary, he does acknowledge that "secular princes are nonetheless not on this account deprived of their great rights" (II, iv, 1). Nonetheless, Aegidius would seem to retract some of the independence and worth that Aquinas accorded the secular order.
24. *Monarchy*, III, xvi.
25. Richard Kay, *Dante's Monarchia* (Toronto: Pontifical Institute of Medieval Studies, 1998), 1–38, 62–63, 119 and passim; Anderson, 223ff. According to Reade, the issue facing Dante is, "the State is either secular or a minor department of the Church" (*Medieval History*, 629).
26. Kay, 18–44 passim.
27. *Medieval History*, 632. Ernst H. Kantorowicz, *The King's Two Bodies: A Study in Medieval Political Theology* (Princeton: Princeton University Press, 1957), 457, argues that Dante advocates unprecedented independence of the state from the church and even from Christianity, by splitting the human from the divine.
28. See Anderson, 214; Dante, *Monarchy*, III, xv.
29. *Inf.* I.
30. Ibid., IV, 42–44, 129, 87–88, 99, 148, 131, 54, 61–62.
31. Ibid., V, 36–37, 69–135.
32. Acts of the Apostles 5:1–6; 8:18–24; *Inf.* XIX, 53–58, 69–78.
33. *Purg.* XVIII, 46–48; XXII, 89–92; XXII, 73–74; XXI, 100–102.
34. Ibid., XXVII, 116–119; XXXII, 99–102.
35. *Par.* XI, 38; XXII; XXII, 72–74, 84–95.
36. Ibid., XII, 140–141.
37. Though Catherine was illiterate, the letters she dictated to the powerful of the world and the church, particularly Pope Gregory XI, helped bring about the return of the papacy from Avignon to Rome in 1376. She was declared a "doctor of the church" in 1970.

## Chapter 5

1. The main translations of Machiavelli used are the following: *Discourses on Livy*, trans. Harvey Mansfield and Nathan Tarcov (Chicago: University of Chicago Press, 1996); *The Prince*, trans. and ed. Harvey C. Mansfield (Chicago: University of Chicago Press, 1985); *The Art of War*, trans. and ed. Neal Wood (Indianapolis, IN: Bobbs-Merrill, 1965); *History of Florence and of the Affairs of Italy*, ed. Felix Gilbert (New York: Harper and Row, 1960). The secondary classics are the following: Jacob Burckhardt, *The Civilization of the Renaissance*

*in Italy: An Essay*, 3rd ed. (New York: Phaidon, 1950); Hans Baron, *The Crisis of the Early Italian Renaissance: Civic Humanism and Republican Liberty in an Age of Classicism and Tyranny* (Princeton: Princeton University Press, 1966; orig. publ. 1955); J. G. A. Pocock, *The Machiavellian Moment: Florentine Political Thought and the Atlantic Republican Tradition* (Princeton: Princeton University Press, 1975).

2. Federico Chabod, *Machiavelli and the Renaissance*, intro. A. P. D'Entrèves (New York: Harper and Row, 1965; orig. publ. 1958), xii. By a single stroke of the pen, Croce revised the age-long denunciation of Machiavelli's immorality. Harvey Mansfield, *Machiavelli's Virtue* (Chicago: University of Chicago Press, 1996), 36–49.

3. Felix Gilbert, *Machiavelli and Guicciardini: Politics and History in Sixteenth Century Florence* (Princeton: Princeton University Press, 1965), pt. 2, 88–93, 168; Donald Wilcox, *The Development of Florentine Humanist Historiography in the Fifteenth Century* (Cambridge, MA: Harvard University Press, 1967), 40–55.

4. Quentin Skinner, *The Foundations of Modern Political Thought*, 2 vols. (Cambridge: Cambridge University Press, 1996; orig. publ. 1978), 1:3–65 passim.

5. This description seems to fit the sixteenth-century "Machiavellian moment" per se than the seventeenth- and eighteenth-century Pocock applies it to, where Stoicism, Ciceronianism, and Protestantism are far more relevant. See Pocock, 169, 183, 190, 202, 214, 360, 461.

6. Felix Raab, *The English Face of Machiavelli: A Changing Interpretation, 1500–1700* (London: Routledge/Kegan Paul, 1964), 32, 54, 257.

7. Leo Strauss, *Thoughts on Machiavelli* (Chicago: University of Chicago Press, 1986; orig. publ. 1958), 242.

8. Aquinas, *On Kingship to the King of Cyprus* (Toronto: University of Toronto Press), II, iii; *Giles of Rome's De regimine principum*, ed. Charles F. Briggs (Cambridge: Cambridge University Press, 1994), 146.

9. Christine de Pizan, *The Book of the Body Politik*, trans. and ed. Kate Langdon Forham (Cambridge: Cambridge University Press, 1994), 35ff.

10. Lester Kruger Born, "The Perfect Prince: A Study in Thirteenth- and Fourteenth-Century Ideals," *Speculum* 3 (1928): 501–502.

11. Wilcox, 40–55; Gilbert, chap. 2.

12. Eugene F. Rice, *The Renaissance Idea of Wisdom* (Cambridge, MA: Harvard University Press, 1958), 28, 41.

13. Ibid., 406.

14. This is especially true of the *Discourses on Livy*.

15. Carlo Battisti and Giovanni Alessio, *Dizionario etimologico italiano*, 5 vols. (Firenze: Barbèra, 1950–1957), s.v. *virtù*.

16. Machiavelli, *Discourses*, I, 9, paras. 1, 2, 5; *The Prince*, chaps. 8, 16, 18, 21, 29, 30.

17. J. H. Hexter, "'Il Principe' and lo Stato," *Studies in the Renaissance* 4 (1957): 125, 128–129, 132–134; *Discourses*, I, 9.

18. Guicciardini, *Ricordi, politici e civili*, ed. Raffaele Spongano (Florence: Sansoni, 1951), C6, C117, 78, 114, 115, 23, 45, 55, 18, 25.

19. Machiavelli, *The Art of War*, trans. and ed. Neal Wood (Indianapolis, IN: Bobbs-Merrill, 1965), bk. 1 passim; *Discourses*, II, 17.2.

20. *Discourses*, I, 13.1; cf. I, 56.1.
21. *History of Florence*, V, 1.
22. Machiavelli, *Art of War*, I, 16. Also see Marcia Colish, "Republicanism, Religion and Machiavelli's Savonarolan Moment," *Journal of the History of Ideas* 60 no. 4 (1999): 599–616. Colish argues in support of Machiavelli's generally favorable attitude to religion, reducing his opposition to it to his aversion to the Savonarolan republic (1494–1498). In the same issue of *JHI*, Benedetto Fontana argues that Machiavelli accepted the papacy as a "unique regime" (639–658). But, as Fontana suggests, Machiavelli appreciated religion mostly when it was politically useful, which is our argument as well. As such Machiavelli has a worldly or secular attitude toward religion, which does not mean that it is a hostile one.
23. Pocock, 214, 183.
24. *The Prince*, chap. 7.
25. Pocock, 85, 133; Machiavelli, *The Prince*, chap. 12. The point may be exaggerated, however. Machiavelli indicates many causes for Italy's ruin. See chaps. 24, 26.
26. *The Prince*, chap. 11.
27. For the medieval origins of the doctrine of the "double truth," see Etienne Gilson, *Reason and Revelation in the Middle Ages* (New York: Scribner's, 1938). It does not seem that any European Averroist actually advocated this doctrine. For its Enlightenment use see Harry Payne, *The Philosophes and the People* (New Haven: Yale University Press, 1976).
28. *The Prince*, chap. 8.
29. Ibid., chap. 8.
30. Ibid., chap. 7.
31. *History of Florence*, VIII, chaps. 1, 2.
32. Ibid.
33. *History of Florence*, I, chaps. 3, 12, 13.
34. *Discourses*, III, chap. 1.4.
35. Strauss, 242; Mansfield discusses these examples in his introduction to the *Discourses*, xxxiii–xxxvi.

## Chapter 6

1. Works consulted: *Luther's Works*, ed. Jaroslav Pelikan et al., vols. 1–55 (Saint Louis: Concordia Press; Philadelphia: Fortress Press, 1955–1976), vols. 31–55, hereafter abbreviated *L.W.*; Heiko A. Oberman, *Luther, Man, Between God and the Devil*, trans. Eileen Walliser-Schwarzbart (New Haven: Yale University Press, 1989; orig. publ. 1982); George Hunston Williams, *The Radical Reformation* (Philadelphia: Westminster Press, 1962); Erik H. Erikson, *Young Man Luther: A Study in Psychoanalysis and History* (New York: W. W. Norton, 1962; orig. publ. 1958); John M Headley, *Church, Empire and World* [Brookfield, VT (USA): Ashgate, 1997]; John M. Tonkin, "Luther's Interpretation of Secular Reality," *Journal of Religious History* 6 (1970): 135–150; Steven E. Ozment, *The Reformation in the Cities* (New Haven: Yale University Press, 1975). Charles J. Herber kindly calculated for me the number of miles Luther must have walked.

2. Erikson, 23–40 passim. The term "theonomous" as opposed to "autonomous" is used by my colleague Dewey Wallace to characterize the God-centeredness of the Reformation as opposed to the Renaissance and Enlightenment autonomy of the individual.
3. *L.W.* 31:265; 11:318.
4. I am thinking of the amiable Lutheran pastor from whom I took a course on the Reformation at Boston College in 1963.
5. *L.W.* 36:91, 41.
6. *L.W.* 31:344; 44:127.
7. Headley, "Luther and the Problem of Secularization," 21–37. Headley here focuses on Luther's desacralized view of nature. Oberman, it might be pointed out, emphasizes Luther's diabolized view of his enemies. A count of the use of the term "devil" in the indexes of *Luther's Works* is at times staggering. Cf. vol. 41. In a 1981 lecture on Luther at the George Washington University, Headley stressed the medieval view of God's immanence in nature, emphasizing love as opposed to Luther's desacralized view of the world. Tonkin's article is another excellent treatment. I am emphasizing Luther's secularizing of the papacy and of lay-clerical relations, celibacy, the sacraments, and political authority.
8. *L.W.* 44:177, and in general "The Judgement of Martin Luther on Monastic Vows," *L.W.* 44:251–400. Erikson discounts a lapse in monastic chastity on Luther's part; see 29, 160. Heinrich Denifle, *Luther and Lutherdom*, 2nd ed., trans. Raymund Volz, 1:1, 99–127, 307–324, shows the disingenuousness of Luther's critique of vows.
9. "Estate of Marriage," *L.W.* 45:21, 42; "An Exhortation to the Knights of the Teutonic Order . . . ," 45:150.
10. *L.W.* 54:307; 24:83, 154–155. Cf. "Judgment on Monastic Vows," *L.W.* 44:251–400.
11. Gustaf Wingren, *Luther on Vocation*, trans. Carl C. Rasmussen (Philadelphia: Muhlenberg Press, 1957), 176, 180, 183.
12. *L.W.* 23:269; 22:61, 366, 84; 47:163, 191–192, 177, 277, 264, 268, 267, 269, 280, 270, 279.
13. *L.W.* 49:150; 46:49–55.
14. Engels, in Robert C. Tucker, *The Marx and Engels Reader* (New York: W. W. Norton, 1978; orig. publ. 1972), 60, 685; Michael G. Baylor, ed., trans., *The Radical Reformation* (Cambridge: Cambridge University Press, 1962); Williams, 341–386, 791–814, 846–865.
15. Baylor, 2, 5, 51, 37.
16. Ibid., 262–263.
17. Ibid., 37, 25, 31, 88–89.
18. Ibid., 210, 206.
19. *L.W.* 45:89–92; W. D. J. Cargill Thompson, *The Political Thought of Martin Luther* (New York: Barnes and Noble, 1984), 52.
20. *L.W.* 23:369.
21. Ibid., 45:96, 111, 113, 105, 115.
22. Ibid., 24:123.
23. Ibid., 43:148; 49:422; 48:188.

24. Gattinara, in John M. Headley, *The Emperor and his Chancellor: A Study of the Imperial Chancellery under Gattinara* (Cambridge: Cambridge University Press, 1983), 153; *L.W.* 50:248; 49:275.
25. *L.W.* 6:244; 35:244.
26. Ibid., 41:248; 6:244; 35:5; Headley, *The Emperor*, 115–116; Thompson, 109–111, 49.
27. Gerald Strauss, "The Reformation and its Public in an Age of Orthodoxy," in *The German People and the Reformation*, ed. R. Po-Chia Hsia (Ithaca, NY: Cornell University Press, 1988), 209.
28. Keith Thomas, *Religion and the Decline of Magic: Studies in Popular Beliefs in Sixteenth and Seventeenth Century England* (New York: Scribner's, 1971); Carlo Ginzburg, *The Cheese and the Worms: The Cosmos of a Sixteenth-Century Miller*, trans. John Tedeski (Baltimore: Johns Hopkins University Press, 1997, 1980); Jean Delumeau, *Catholicism from Luther to Voltaire* (Philadelphia: Westminster Press, 1977; orig. publ. 1971); John Bossy, "Holiness and Society," *Past and Present* 75, no. 1 (1977), 119–137; Gabriel le Bras, *Etudes de sociologie religieuse* (Paris: Presses Universitaires de France, 1955).
29. Richard H. Popkin, *The History of Scepticism: From Savonarola to Bayle* (Oxford University Press, 2003), chap. 1; *L.W.* 33:passim. Luther uses the term "schismatics" to include Catholics, especially the papacy, Jews, Turks, and radicals.
30. Popkin, chap. 2.
31. Ibid., 48, chap. 3; Donald Frame, *Montaigne: A Biography* (New York: Harcourt Brace and World, 1965), chap. 10.
32. Chaunu, *La civilization de l'Europe classique* (Paris, 1970), 398.

## Chapter 7

1. Main works used: John Locke, *Two Tracts of Government*, ed. Philip Abrams (Cambridge: Cambridge University Press, 1967); John C. Biddle, trans., ed., "John Locke's 'Essay on Infallibility': Introduction, Text and Translation," *Journal of Church and State* 19 (1977): 301–327; Locke, *Essays on the Law of Nature: The Latin Text with English Translation*, ed. Wolfgang von Leyden (Oxford: Clarendon Press, 1954); *The Correspondence of John Locke*, 8 vols., ed. E. S. De Beer (Oxford: Clarendon Press, 1976–1989)—all references to this work will be hereafter abbreviated *Corr.*, followed by the volume number (in roman) and the letter number (in Arabic); "An Essay Concerning Toleration," in *The Life of John Locke*, 2 vols., ed. Henry Richard Fox Bourne (Aalen: Scientia Verlag, 1969; orig. publ. 1876), 1:174–194; *A Letter Concerning Toleration*, ed. James H. Tully (Indianapolis, IN: Hackett, 1983; orig. publ. 1689); Proast, *The Argument of the "Letter Concerning Toleration," Briefly Consider'd and Answer'd* (London: Theatre for George West; Oxford: Henry Clements Booksellers, 1690); Locke, *Works*, 10 vols. (London: William Tegg, 1823), (references to all four letters concerning toleration except the first are to vol. 6 of Locke's *Works*); I. T. Ramsey, ed., *The Reasonableness of Christianity with A Discourse of Miracles and Part of a Third Letter Concerning Toleration* (Stanford: Stanford University Press, [1958]; Maurice Cranston, *John Locke: A Biography* (New York: Macmillan, 1957) 1, 3, 16–21, 27, 29, 39, 73, 75, 96

on Locke's early life. Cranston describes Latitudinarianism as "essentially an Anglican movement" (126), but, according to Dewey Wallace, a strong strain of Dissenting Latitudinarianism can be found in Richard Baxter (1615–1691). John Marshall, *John Locke: Resistance, Religion and Responsibility* (Cambridge: Cambridge University Press, 1994), 51–52. This book is an excellent treatment of the evolution of Locke's religious thought in relation to his historical context and to the political challenges he faced. The interpretation that follows below differs from Marshall's in that I interpret Locke as the pivotal, modern secularist rather than as a religious and political dissenter. Nicholas Wolterstorff, *John Locke and the Ethics of Belief* (Cambridge: Cambridge University Press, 1996), stresses the importance to Locke's thought of the Reformation, which constituted a "cataclysm. Locke's epistemology was addressed to that cataclysm"(246). The medieval legacy reached him in a fractured state, and he was confronted throughout his life with the need to counter "opinion" (*doxa*) with philosophical proofs, particularly the ones concerning the existence of God and the relation of reason to faith; see Locke's *Essay Concerning Human Understanding* ed. Peter Nidditch (Oxford: Clarendon, 1975), IV, x, i; IV, xviii, xx.10). On continental skepticism and its links to the Reformation, see Richard Popkin, *A History of Skepticism: From Savonarola to Bayle* (Oxford: Oxford University Press, 2003; orig. publ. 1979), chap. 1.
2. Locke, *Two Tracts*, 118–120, 210. Neither the sects of the civil wars nor Cromwell embraced toleration; cf. Marshall, 59n. It would be interesting to know by what steps Locke corrected his supposed Puritan upbringing to become so critical of dissenting idiosyncrasies.
3. Locke, *Two Tracts*, 124; Biddle, 301–327.
4. *Corr.* I, 30.
5. Ibid., 180, cf. 75.
6. Locke, *Essays on the Law of Nature*; "First Tract," 184; Michael Oakeshott, "Locke: The Theological Vision," in *Morality and Politics in Modern Europe: The Harvard Lectures*, ed. Shirley Robin Letwin (New Haven, CT: Yale University Press, 1993), 54–55. "Natural Law is the universal Moral Law knowable by reason. Divine Law is revealed by God as with the Ten Commandments which reinforce the Moral Law."
7. Locke, "First Tract," 159.
8. Ibid., 160–161.
9. Ibid., 164. The word "ancient" here could refer to the "Catholic" Anglican church or to the pre-Reformation church.
10. Cited ibid., 140n.
11. "Second Tract," 218, 229. This tract was written in Latin and translated by the editor Philip Abrams. As we have seen Chapter 5, papal infallibility was claimed by the bishop of Rome at least as early as Gregory VII's *Dictatus Papae* of 1090. The official definition of the doctrine, limited to *ex cathedra* statements on faith and morals, was not decreed until 1870. On *The Reasonability of Christianity*, see n. 54 below.
12. Locke, "An Essay Concerning Toleration," 1:174–194, and "Essay on Infallibility," 321, 323, 317; Marshall has found four drafts of the *Essay concerning Toleration* and with considerable change through them (51–52, 58, 61–64). The one cited here, for instance, disapproves of governmental interference in

speculative theological opinions or worship (although enforcement of the *performance* of divine worship is maintained). No arrests should be made for (private?) immorality and so on. It is clear that Locke found these issues unsettling but was moving toward a more secular solution of these problems.

13. "Essay on Toleration," 182; Marshall, 51–52, 58, 72.
14. "Essay on Toleration," 186, 179.
15. Ibid., 177. One thinks ahead here to Voltaire's *Lettres philosophiques* (1732), in which a Christian, a Jew, and a Muslim trade peacefully on the stock exchange together.
16. "Essay on Toleration," 179.
17. Ibid., 176, 183. Oakeshott employs the term "umpire" to describe the central figure of Locke's political and moral theory (47–58).
18. "Essay on Toleration," 183–184, 187–188.
19. "Essay on Infallibility," 317; *Essay on Toleration*, 183; cf. Marshall, 67–68, 53–54, 76. Marshall dates a volte-face in Locke's views on toleration at 1667. Neither the civil wars sects nor Cromwell espoused toleration, ibid., 59n. Roman Catholics, as well as Quakers, were excluded from consideration for tolerance in the four drafts of Locke's *Essay on Toleration*, 53. This is the one tenet (exclusion of Catholics from civil rights) that Locke did not change through his life. He considered any church that would impose on others its rule on "indifferent things" to be the "Antichrist" (Marshall, 21).
20. "Essay on Infallibility," 325, 327.
21. Locke owned a copy of Machiavelli's works; see John Harrison and Peter Laslett, *The Library of John Locke* (New York: Oxford University Press, 1965), item 1848. Locke had accepted the Test Act of 1763 by this time (Marshall, 76–84).
22. "Essay on Infallibility," 317; "First Tract," 131, 174.
23. Theodore K. Rabb, *The Struggle for Stability in Early Modern Europe* (New York: Oxford University Press, 1975).
24. Cranston, 184–230 passim. Ashley was identified with Arminian (supporters of free will vs. predestination), Latitudinarian, and anti-Stuart policies. Cf. Marshall, 70, 119, 122–123, 221–222, 231, 257, 262–263, 352–357, 359; Richard Ashcraft, *Revolutionary Politics: Locke's Two Treatises of Government* (Princeton: Princeton University Press, 1986), 137–138, 80–81, 84, 121, 327. For the Test Acts see *English Historical Documents*, vol. 8, *1660–1714*, ed. Andrew Browning (New York: Oxford University Press, 1953), 391–392.
25. Cranston, 260, 321, 246, chap. 20 passim; Marshall, 90–95.
26. Cranston, 269–291, 293.
27. *Corr*. III, 963, 1058, 1215; V, 1708, 1823, 1878; III, 1131. Remonstrants were liberal Calvinists who stressed conditional rather than absolute predestination and, like the Armenians, accepted a degree of free will.
28. Cranston, 75–77. On Isaac Newton's religious scholarship, see Richard S. Westfall, "Isaac Newton's *Theologia gentilis origines philosophicae*," in *The Secular Mind: Transformation of Faith in Modern Europe; Essays Presented to Franklin L. Baumer*, ed. W. Warren Wagar (New York: Holmes and Meier, 1982), 15–34. Newton was a far more adamant anti-Trinitarian than Locke. The scientist believed that "the necessary truths had been learned from nature in the Age of Noah and . . . [he] effective[ly] denied that Christ revealed any truth" (ibid., 30).

29. Cranston, 309–311; *Corr.* III, 1147n.
30. *Corr.* III, 1182.
31. Locke, *Letter Concerning Toleration*, 23. This is William Popple's translation of the *Epistola de Tolerantia* that had been published in Gouda earlier that year. Locke had long believed that morality was more important than doctrine, that it was demonstrable by reason and that the state had the right to enforce it (*Essay Concerning Human Understanding*, bk. IV), that it was knowable by reason from the law of nature, and that it should be a transcript of the Gospels. But in the last decade of his life, he came to believe that no moral code was possible. Objective morality had originallly been for Locke a fall-back position for the lack of doctrinal agreement. Cf. Marshall, 200, 387, 435–437, 447.
32. *A Letter Concerning Toleration*, 37–38; Marshall, 65.
33. *A Letter Concerning Toleration*, 38.
34. Ibid., 37. It seemed that Locke thought original sin could be mended by natural means, namely education, rather than by supernatural (sacramental) means. Cf. Marshall, 340–343. Clearly this is an important detail in Locke's secularism.
35. The expression "orthodox to oneself" could mean that there is no such thing as orthodoxy in an absolute sense, or it could mean that every man considers himself in the right and that no one considers himself heretical. The statement pertains to the subjective vs. the objective dimension of Locke's epistemology and evokes William of Ockham's nominalism. Cf. "Most of those [ideas] of sensation being in the mind [are] no more the likeness of something existing without us, than the names, that stand for them are the likeness of *ideas*" (*Essay Concerning Human Understanding*, II, viii, 7) and "We should have a great many fewer disputes in the world, if words were taken for what they are, the signs of our *ideas only,* and not for things themselves" (ibid., III, x, 15). Thus, because we think we are orthodox does not mean that we are, nor, in Locke's view, that anyone can force us to be so (according to one's own idea of orthodoxy). The simplest thing would seem to be to abandon the idea of orthodoxy altogether, which is what he does in his several *Letter*[s] *Concerning Toleration* (1689– ). See my "Enlightenment Anticipations of Postmodernist Epistemology," *Poznań: Studies in the Philosophy of the Sciences and the Humanities* 58 (1997): 105–121.
36. *A Letter Concerning Toleration*, 26. Locke goes so far as to say, "There is absolutely no such Thing, under the Gospel, as a Christian Commonwealth." (ibid., 44). Augustine had intimated the same. See Chapter 2.
37. Ibid., 26, 42, 45, 54.
38. Ibid., 51; Pierre Bayle, *Historical and Critical Dictionary: Selections*, ed. Richard Popkin (Indianapolis, IN: Hackett Press, 1991; orig. publ. 1965), 399–408. Locke, *Works*, 116. Citations to the second, third, and fourth letters on toleration are contained in *Works*, vol. 6.
39. Cranston, 331; Proast, *The Argument of the Letter Concerning Toleration*.
40. Proast, 5,15. Proast never specifies what kind of coercion he would employ.
41. *Second Letter*, 90.
42. *Third Letter*, 182.
43. Locke had doubts about the Trinity through most of his life, but it was not until the publication of *The Reasonableness of Christianity* in 1695 that they

became known and were attacked for Socinianism or Unitarianism, notably by Bishop Edward Stillingfleet. See Marshall, 138, 140–141, 154. The same can be said about his belief in original sin, which he simply omitted from discussion in *The Reasonableness.*

44. *Second Letter,* 64, 110, 132, 65. On Locke's views on original sin, see the very nuanced study by W. M. Spellman, *John Locke and the Problem of Depravity* (Oxford: Clarendon Press, 1988).
45. *Fourth Letter,* 569; *Second Letter,* 76, 64, 83; *Third Letter,* 397, 405, 517; Reasonableness of Christianity (1695) in Works, 7:4–178.
46. *Third Letter,* 544, 188; *Second Letter,* 65.
47. *Third Letter,* 402, 252, 370.
48. *Second* and *Third Letter,* 72, 257, 76–77, 279, 134.
49. *Essay,* II, xxi; *Second Letter,* 74–76, 118.
50. *Second Letter,* 74, 117; *Third Letter,* 225, 227, 209, 212, 170, 194, 195.
51. *Third Letter,* 188. On this point see Oakeshott, and Reinhard Brandt, "The Center-Point of Locke's Philosophy: An Outline," *Locke Newsletter,* no. 14 (1983), 27–33; *Second Letter,* 101. The *ekklesia* (the Greek word for "assembly") designated the church for the early Christians.
52. See the well-edited, abridged version by I. T. Ramsey, *The Reasonableness of Christianity.*
53. *Corr.* VIII, 3450; *Vindication of the Reasonableness of Christianity* is in *Works* 7:159–190; *A Second Vindication of the Reasonableness of Christianity* follows in *Works* 7:191–424. See Marshall (352–353, 376) on submitting Revelation to reason, bringing it under individual examination. The irony is that this theological boldness was accompanied by an epistemological pessimism. The afterlife could not be proved, for instance, but Locke believed it (without the resurrection of bodies), ibid., 286–288, 399.
54. *Corr.* VIII, 3328; emphasis added.
55. *Corr.* VI, 2207; V, 1880; VI, 2240; VI, 2601; VIII, 3456; VII, 2881; VI, 2631; VII, 2935; VIII, 3339. The short list of "heretics" put to death by the English government that Locke obtained would be considerably augmented by those of Catholic priests, Jesuits in particular, and their hosts, in the reign of Elizabeth and James I. See John Gerard, *The Autobiography of a Hunted Priest,* ed. Philip Caraman, intro. Graham Greene (New York: Doubleday, 1955; orig. publ. 1952).
56. The issue of matter thinking is discussed in the *Essay Concerning Human Understanding,* IV, iii, 6; *Corr.* VIII, 3342, 3351.
57. Locke did not see how a belief in infallibility could be compatible with a belief in toleration. It could be found in the life of someone like his contemporary Francis de Sales, bishop of Geneva. On septic fideism, see Popkin, *History of Skepticism,* passim. Popkin's first chapter considers the causal connection between the Protestant Reformation and the rise of skepticism. Locke's intellectual biography bears out the thesis.
58. Pierre Bayle was the editor and publisher of the *Nouvelles de la République des Lettres* (1684–1687), wherein the term likely originated. David Berman, "Deism, Immortality and the Art of Theological Lying," in *Deism, Masonry, and the Enlightenment: Essays Honoring Alfred Owen Aldridge,* ed. J. A. Leo Lemay (Newark, NJ: University of Delaware Press, 1987), 61–78. Berman

argues that deist protestations of belief in immortality are not to be trusted because deists usually argued against such beliefs. Then to protect themselves (from the public as much as from the authorities), they paid lip service to "revelation." Locke is shown to be a fideist in his *Second Reply to Stillingfleet* [Bishop of Worcester] (1699; 1722, *Works* IV), thus protecting by faith what he could not affirm by reason. Berman considers the "fideism" of Hume to be wholly disingenuous (70–72). Locke can be read as undermining the traditional political and intellectual structures of Christianity (the confessional state and scholasticism or Aristotelianism), while championing an individualist position in biblical exegesis. This individualism is one of the two *termini* of Protestantism, a legacy of Luther's antischolastic nominalism and his solafideism. It is paralleled by some continental skeptics, such as Pierre Bayle. The second *terminus* is the antirational evangelicalism, especially Methodism, which rocked the Atlantic world at the end of the eighteenth century. I believe Locke is the consummate Protestant skeptic, much as Peter Gay believes Hume to be the consummate modern pagan.

## Chapter 8

1. For the materialist interpretation of Rousseau, see Mark Hulliung, *The Autocritique of the Enlightenment: Rousseau and the Philosophes* (Cambridge, MA: Harvard University Press, 1994).
2. The works used in this chapter (by Rousseau, unless otherwise indicated) are abbreviated as follows: *O.C.*—*Oeuvres complètes*, ed. Bernard Gagnebin and Marcel Raymond, 5 vols. (Paris: Bibliothèque de la Pleiade, Gallimard, 1959–1995); *Beaumont*—[*Lettre*] *à M. Christophe de Beaumont*, ed. Henri Gouhier, *O.C.* 4; *Confessions*-Garnier—*Les confessions*, ed. A. Van Bever, 3 vols. (Paris: Garnier, n.d.); *Confessions*-Pleiade—*Les confessions*, ed. Bernard Gagnebin and Marcel Raymond, *O.C.* 1; Cranston I—Maurice Cranston, *Jean-Jacques: The Early Life and Works of Jean-Jacques Rousseau, 1712–1754* (Chicago: University of Chicago Press, 1982); Cranston II—Cranston, *The Noble Savage, Jean-Jacques Rousseau, 1754–1762* (Chicago: University of Chicago Press, 1991); Cranston III—Cranston, *The Solitary Self: Jean-Jacques Rousseau in Exile and Adversity* (Chicago: University of Chicago Press, 1997); *Emile*-Pleiade—*Emile, ou de l'éducation*, ed. Charles Wirz and Pierre Burgelin, *O.C.* 4; *Emile*-Garnier—*Emile, ou l'éducation*, ed. Michel Launay (Paris: Garnier-Flammarion, 1966); *First Discourse*—*Discourse on the Sciences and Arts*; *Second Discourse*—*Discourse on the Origin and Foundations of Inequality*, in Rousseau, *First and Second Discourses*, trans. Judith R. Masters, ed. and intro. Roger D. Masters (New York: St. Martin's Press, 1964); *Julie*—*Julie, ou la Nouvelle Héloïse*, ed. Henri Coulet and Bernard Guyon, *O.C.* 2; *Montagne*—*Lettres écrites de la Montagne*, ed. Jean Daniel Candaux, *O.C.* 3; "Notes sur 'De l'Esprit' d'Helvétius," ed. H. Gouhier, *O.C.* 4; Ronald Grimsley, ed., *Rousseau, Religious Writings* (Oxford: Clarendon Press, 1970); *Masson*—Pierre Maurice Masson, *La religion de J. J. Rousseau*, 3 vols. (Paris: Hachette, 1916); *Rêveries*—*Rêveries du promeneur solitaire*, ed. Marcel Raymond, *O.C.* 1; *S.C.*—*Social Contract*, trans. and ed. Maurice Cranston (Harmondsworth, UK: Penguin, 1986; orig. publ. 1968) [Citations to this last work will be to book and chapter]; "Voltaire"—"Lettre de J. J. Rousseau à

Voltaire," *O.C.* 4; "Malesherbes"—"Quatre Lettres à. M. le President de Malesherbes," *O.C.* 4. For revolutionary Rousseauiana, see André Monglond, *La France révolutionnaire*, 10 vols. (Grenoble: Arthaud, 1930–1978), vols. 1–4, Cranston I; Cranston II; Cranston III.
3. Cranston I, 41; *Confessions*-Garnier 1:78.
4. The best source is Cranston's biography, chaps. 1–5, passim, 174.
5. "Malesherbes," January 12, 1762, 1135–1136. The revelation or "illumination" merits comparison with Descartes' dream.
6. *First Discourse*, 61n and passim.
7. *Second Discourse*, 225. Rousseau indicts European Christian society, not that of non-Europeans. The concept of the noble savage was revolutionary, in that natives exchanged places with Europeans in the monopoly of virtue. On the sixteenth-century controversies over the humanity of savages, see Anthony Pagden, *The Fall of Natural Man: The American Indian and the Origins of Comparative Ethnology* (Cambridge: Cambridge University Press, 1982); John H. Zammito, *Kant, Herder and the Birth of Anthropology* (Chicago: University of Chicago Press, 2002); and last, but not least, Diderot and D'Alembert's *Encylopédie*, s.v. "Hottentot," which lists the large number of voyagers (whom Rousseau could have read) who had commented on this people. For perhaps the most thorough and scholarly treatment of Rousseau's social-political thought, see Robert Wokler, *The Social Thought of J. J. Rousseau* (New York: Garland, 1987).
8. *Rêveries*, 1075.
9. *Confessions*-Garnier 1:293; *Rêveries*, 1012, 1779n. In his second letter to Malesherbes, describing his youth, Rousseau writes, "I was busy because I was crazy. To the extent that my eyes were opened, I changed tastes, friends, projects and in all these changes, I always lost time and my effort because I was always looking for something that did not exist" ("Malesherbes," 1134).
10. *Confessions*-Garnier 2:254–255. I have emended my translation of this passage with that of "The Confessions and Correspondence Including the Letters to Malesherbes," in *The Collected Writings of Rousseau*, ed. Christopher Kelly et al. (Hanover, NH: University Press of New England, 1990), 5:346–347. Trio is a card game.
11. *Confessions*-Garnier 2:179–180; Cranston I, 244–246, supports Rousseau's paternity rather than Thérèse's putative lovers.
12. *Rêveries*, 1015.
13. *Beaumont*, 935–936.
14. On hierarchy and discipline in the old regime, see Charles Loyseau, *A Treatise on Orders*, excerpt translated in Keith Michael Baker, *The Old Regime and the French Revolution* (Chicago: University of Chicago Press, 1987), 13–31, and Philippe Aries, *Centuries of Childhood: A Social History of Family Life* (New York: Random House, 1962), 46; *Emile*, passim; *S.C.* I, i; Immanuel Kant, "What is Enlightenment?" in *On History*, trans. and ed. Lewis White Beck et al. (Indianapolis, IN: Bobbs-Merrill, 1963), 3.
15. *Beaumont*, 935–936; Cranston I, 265–266. "Obey only oneself" is half the formula Rousseau gives in *S.C.* I, 6, for the "social pact."
16. *Emile*-Pleiade, 245; *S.C.* I, i.

17. *S.C.* I, 8; II, 6.
18. Kant, *Critique of Pure Reason*, trans. Norman Kemp Smith (New York: St. Martin's Press, 1965; orig. publ. 1929), 30; Ronald Grimsley, *Rousseau and the Religious Quest* (Oxford: Clarendon Press, 1968), 137.
19. *Emile*-Garnier, 335, 395, 388; Voltaire, *Philosophical Dictionary*, passim; *Emile*-Pleiade, 628, 628n, 621; *S.C.* IV, viii.
20. *Beaumont*, 937–938, 1013; *Emile*-Pleiade, 624, 1583 ed.n.; Malebranche, *De la recherche de la vérité*, 2 vols. (Paris, 1688), 2:367–368.
21. *Emile*-Pleiade, 613–614, 751, 960, 1013, 1577–1578 ed.n; *Beaumont*, 999, 1013; *Emile*-Garnier, 397; *O.C.* 1:728n, 1649.
22. *Confessions*-Garnier 1:62–63, 222–225; 3:110, 114; 1:122–124. On the abbés Gatier and Gaime of Savoy, who were the models of the Vicar Savoyard, see *Confessions*-Garnier 1:161. On Rousseau's favorable impression of a Jesuit, see *Confessions*-Garnier, 2:161.
23. Grimsley, *Religious Writings*, 71; *Emile*-Garnier, 345. It will be remembered from the last chapter that Locke's preoccupation was less the global penetration of the Gospel than the conflicting global interpretations of it.
24. *Montagne*, 728n, 768.
25. Cranston III, 2, 6; Rousseau defended himself saying, "I have religion, my friend, and so much . . . that I do not believe there is any man in the world who needs it as much as I do." He defends the Genevan pastors against D'Alembert's charge that they were Socinians and expresses the belief that it was not necessary to adhere to a church's abstruse doctrine in order to be saved. Grimsley, *Religious Writings*, 68, 375–380.
26. *Mandement de Monseigneur l'Archevêque de Paris, portant condamnation d'un livre qui a pour titre "Emile ou l'education," par J. J. Rousseau* (Paris: Simon, 1766); *Beaumont*, 1003, 1005, 927, 950, 953, 991, 995.
27. Six years earlier he had defended the Genevan pastors against D'Alembert's *Encyclopedia* article "Geneva," in his *Lettre à D'Alembert sur les spectâcles* (1758); Cranston III, passim.
28. *Emile*-Pleiade, 570.
29. *Emile*-Garnier, 350, 362 and passim, bk 4; *O.C.* 4:943.
30. Ibid., 581; G. Leibniz, *Theodicy, Essays on the Goodness of God, the Freedom of Man, and the Origin of Evil*, trans. E. M. Huggard, ed. Austin Farrer (London: Routledge/Kegan Paul, 1952); Samuel Clarke, *A Discourse concerning the Being and Attributes of God, the Obligations of Natural Religion and the Truths and Certainty of the Christian Religion*, in *Works*, 4 vols. (London: John and Paul Hampton, 1738) 2:passim; *The Leibniz-Clarke Correspondence*, ed. H. G. Alexander (Manchester: Manchester University Press, 1956); Matthew Tindal, *Christianity as Old as the Creation* (New York: Garland, 1978, orig. publ. 1730), 58, 70, 246–247, 249.
31. *S.C.* IV, 8.
32. Ibid.
33. Ibid.; *Montagne*, 772.
34. *S.C.* IV, 8. On the concept of "ancient liberty," see Stephen Holmes, *Benjamin Constant and the Making of Modern Liberty* (New Haven: Yale University Press), 1984, pt. 1.

35. *S.C.* IV, 8.
36. Georges Lefebvre, *The French Revolution*, trans., Elizabeth Moss Evanston, 2 vols. (New York: Columbia University Press, 1964; orig. publ. 1962), vol. 1, *From Its Origins to 1793*, 244. Carol Blum, *Rousseau and the Republic of Virtue: The Legacy of Politics in the French Revolution* (Ithaca, NY: Cornell University Press, 1986), does not examine the relevance of the last chapter of the *Social Contract* on the Revolutionary festivals, particularly the cult of the Supreme Being of June 8, 1794.
37. *S.C.* IV, 8.
38. Ibid.; Keith Michael Baker, *Inventing the French Revolution: Essays on Political Culture in the Eighteenth Century* (Cambridge: Cambridge University Press, 1990), pt. 3; Emmet Kennedy, "The French Revolution and the Genesis of a Religion of Man, 1760–1885," in *Modernity and Religion*, ed. Ralph McInerny (Notre Dame, IN: Notre Dame University Press, 1994), 61–88.
39. Mona Ozouf, "Le Panthéon, l'Ecole normale des morts," in *Les Lieux de mémoire*, 3 vols., ed. Pierre Nora, vol. 1, *La République* (Paris: Gallimard, 1984–1986), 139–166. For Rousseau and the counterrevolutionaries, see the studies by Roger Barny.

## Chapter 9

1. This was most evident in Condorcet's *Esquisse d'un tableau historique des progrès de l'esprit humain* (Paris: Agasse, an III [1795]); Martin Staum, *Minerva's Message: Stabilizing the French Revolution* (Montreal: McGill, 1996).
2. The Library of Congress holds over 1,500 works by and about Kant. I have basically restricted myself to Kant's works in English translation. They are *Critique of Pure Reason* [1787], 2nd ed., trans. Norman Kemp Smith (New York: St. Martin's Press, 1965; orig. publ. 1929), hereafter *CPRb*; *Critique of Practical Reason* [1788], trans. Lewis White Beck (Indianapolis, IN: Bobbs-Merrill, 1956), hereafter *CPraR*; *Religion within the Boundaries of Mere Reason* [1793], *and Other Writings*, trans. Allen Wood et al. (New York: Cambridge University Press, 1998), hereafter *R.B.*; *Prolegomena to any Future Metaphysics* [1783], trans. Paul Carus and James W.Ellington (Indianapolis, IN: Hackett, 1977; orig. publ. 1902); On History [1784], trans. Lewis White Beck et al. (Indianapolis, IN: Bobbs-Merrill, 1963; orig. publ. 1957); *Foundations of the Metaphysics of Morals*, ed. Lewis White Beck (Indianapolis, IN: Bobbs-Merrill, 1997; orig. publ. 1959), hereafter *Foundations*; *Correspondence*, trans. and ed. Arnulf Zweig (Cambridge: Cambridge University Press, 1999), hereafter *Corr*. Additionally, the following have been helpful: *A Kant Dictionary* (London: Blackwell, 1995), hereafter K.D.; Manfred Kuehn, *Kant: A Biography* (Cambridge: Cambridge University Press, 2001); John Zammito, *Kant, Herder and the Birth of Anthropology* (Chicago: University of Chicago Press, 2002).
3. *Corr.*, 1–10.
4. Ibid., 9.
5. The term "dogmatist" occurs frequently in Kant's work; cf. *CPRb*, 32–33. Also cf. K.D., 162–163. Dogmatism can be likened to the French Encyclopedists' use of the term esprit de *système*, which refers disparagingly to Cartesianism.

Kant most likely refers to the residue of scholasticism in German philosophy. See Hume, *Enquiry Concerning the Principles of Human Understanding* [1746], sec. VII and passim; Peter Gay, *The Enlightenment: An Interpretation: The Rise of Modern Paganism* (New York: Knopf, –1966), 67.
6. Locke, *Essay Concerning Human Understanding*, ed. Peter Nidditch (Oxford: Clarendon Press, 1975), III, x, 23; *CPRb*, 24–27.
7. *CPRb*, intro., pref., 21–33; ibid., intro., chaps. 6 and 7.
8. *CPRb*, 113–114, 128, 143, 160. The example is mine.
9. *K.D.*, 362–365, 118–121.
10. *K.D.*, 102–106, 118–121, 262–266; *CPRb*, div. I, bk. 1; div. II, bk. 2, chap. 2, 409, 415.
11. *CPRb*, 328, 297–570 passim.
12. Ibid., 396–421; K.D., 75–78.
13. Cf. Condorcet, *Cinq mémoires sur l'instruction publique*, ed. Charles Coutel and Catherine Küntzler (Paris: Flammarion, 1994), 88–93, 96. See Chapter 11.
14. *CPRb*, 29.
15. *Foundations*, 18.
16. Ibid., 56–57, 47, 72, 111, 23–24; *CPraR*, 30.
17. *Foundations*, 81.
18. Kuehn, 361–362.
19. Ibid., 339, 342, 363–364.
20. *CPRb*, 325n; *CpraR*, 100.
21. R. R. Palmer, *The Age of the Democratic Revolution*, 2 vols. (Princeton: Princeton University Press, 1959, 1964), 2:127 and n. Cf. also my *A Cultural History of the French Revolution* (New Haven: Yale University Press, 1989), 345, 338–353; Donald Greer, *The Incidence of the Terror during the French Revolution* (Cambridge, MA, 1935), 165.
22. *R.B.*, 94–102, 40, 114–122, 128, 135, 136, 143–155. Kant sets historical revelation at variance with natural religion, which unlike the former is a religion of reason.
23. Kant, "Idea for a Universal History from a Cosmopolitan Point of View," in *On History*, 11–12.
24. Ibid, 20. According to J. G. A. Pocock, in conversation with the author, Gibbon's attitude toward the French Revolution was the same as Burke's. One might add that Gibbon feared its relapse into barbarism, from which it had recently emerged; Burke feared a wholly new kind of barbarism.
25. François Azouvi and Dominique Bourel, *De Königsburg à Paris: La réception de Kant en France* (1788–1804) (Paris: J. Vrin, 1991), 78–79. For a recent analysis of Kant and Tracy see Michèle Crampe-Casnabet, "Du sytème à la Méthode, Tracy observateur lointain de Kant," *Corpus, Revue de Philosophie*, nos. 26/27 (1994): 75–89. Cf. the author's *A Philosophe in the Age of Revolution: Destutt de Tracy and the Origins of "Ideology"* (Philadelphia: American Philosophical Society, 1978), ch. 4.
26. Tracy to Droz, 27 fructidor an X [September 14, 1802], *Corpus: Revue philosophique*, nos. 26/27 (1994), 202.
27. Kennedy, *A Philosophe*, 46ff.
28. Azouvi and Bourel, 81–98, 103, 105–111.
29. Kennedy, *A Philosophe*, chap. 4.

## Chapter 10

1. Works cited in this chapter are Hegel's, unless otherwise indicated, and are abbreviated as below. Other works are cited by the author's last name and a short title where necessary. Hegel, *Aesthetics, Lectures on Fine Art*, trans. T. M. Knox, 2 vols. (Oxford: Clarendon Press, 1975), hereafter *Aesthetics*; *Early Theological Writings*, trans. T. M. Knox, ed. Richard Kroner (Chicago: University of Chicago Press, 1948), hereafter *E.T.W.*; *Encyclopedia of the Philosophical Sciences in Outline and Critical Writings*, trans, A. V. Miller and Steven A. Taubeneck, ed. Ernst Behler (New York: Continuum, 1990), hereafter *Encyclopedia*; *Lectures on the History of Philosophy*, trans. E. S. Haldane and Francis H. Simson, 3 vols. (London: Routledge/Kegan Paul, 1955; orig. publ. 1896), vol. 3, hereafter *History of Philosophy*; *Lectures of the Philosophy of Religion*, trans., R. F. Brown et al., ed. Peter C. Hodgson, 3 vols. (Berkeley: University of California Press, 1984–1987), hereafter *Philosophy of Religion*; *The Letters*, trans. Clark Butler and Christiane Seiler, ed. Clark Butler (Bloomington, IN: Indiana University Press, 1984), hereafter *Letters*; *Phenomenology of Spirit*, trans. A. V. Miller, ed. A. J. Findlay (Oxford: Oxford University Press, 1977), hereafter *Phenomenology*; *The Philosophy of History*, trans. and ed. C. J. Friedrich (New York: Dover Press, 1956), hereafter *Philosophy of History*; *Philosophy of Right*, trans. and ed. T. M. Knox (rpt. 1978; London: Oxford University Press, 1952), hereafter *Philosophy of Right*; *Three Essays, 1792–1795: The Tübingen Essay, Berne Fragment, The Life of Jesus*, trans. and ed. Peter Fuss and John Dobbins (Notre Dame, IN: University of Notre Dame Press, 1984), hereafter *Three Essays*. The following are the secondary sources: Emil L. Fackenheim, *The Religious Dimension in Hegel's Thought* (Boston: Beacon Press, 1970; orig. publ. 1967); George Armstrong Kelly, *Idealism, Politics and History: Sources of Hegelian Thought* (Cambridge: Cambridge University Press, 1969); Walter Kaufmann, *Hegel: A Reinterpretation* (New York: Doubleday, 1966); Alexandre Kojève, *Introduction à l'étude de Hegel* (Paris: Gallimard, 1947); H. S. Harris, *Hegel's Development*, vol. 1, *Toward the Sunlight, 1770–1801* (Oxford: Clarendon Press, 1972); Harris, *Hegel: Phenomenology and System* (Indianapolis, IN: Hackett, 1995); Emmet Kennedy, *A Philosophe in the Age of Revolution: Destutt de Tracy and the Origin of "Ideology"* (Philadelphia: American Philosophical Society, 1978); W. T. Stace, *The Philosophy of Hegel: A Systematic Exposition* (rpt. 1955; New York: Dover, n.d. [1923–1924]); Jean Wahl, *Le Malheur de la conscience dans la philosophie de Hegel* (New York: Garland, 1984; orig. publ. 1929, Paris).
2. Harris, *Hegel's Development*, 47–56, 1, 2, 47n, 60–64.
3. Ibid., xx, 59, 60n, 91–95, 108–109, xxii, 136, 143, 145, 236, 237n, xviii, xxvi, 123–124, 142–143n, 190–193 and passim; Hegel to Niethammer, September 27, 1814, in *Letters*, 302; Harris, *Hegel: Phenomenology*, vii, 7, 3–5, 14–17, 74–75; Hegel, *Three Essays*, 104–105; *E.T.W.*, 182–301 passim.
4. *Three Essays*, 30–168; *E.T.W.*, 67–330; Harris, *Hegel: Phenomenology*, 10.
5. Harris, *Hegel's Development*, xxxi; *Letters*, 100.
6. Harris, *Hegel: Phenomenology*, 11–12; *Letters*, 125–233, 316, 423–439, 608–664, esp. 615; *Aesthetics*, 2:825.
7. Harris, *Hegel: Phenomenology*, 13–21; *Letters*, passim.

# Notes / 253

8. See Nikolaus von Thaden's critique of Hegel's *Philosophy of Right* [*Law*] in *Letters*, 463, 532; *Philosophy of Right*, 160 (no. 260).
9. Harris, *Hegel's Development*, xvi; Hegel to Ravenstein, May 10, 1829, *Letters*, 543: "I do not know how to satisfy your wish to obtain a copy of a notebook of my lectures on the science of religion. You will more easily obtain this through connections with students among whom such notebooks are circulating, though they do so without my knowledge and—according to the few I have had occasion to see—not exactly always to my satisfaction." Cf. 565, 640. On the constitution of the texts of these lectures, see Kaufmann, 223–224; on the constitution of the critical edition of the *Philosophy of Religion*, see the introductions to them by Peter C. Hodgson.
10. Kaufmann, 368, 95, 99, 100, 220, 368; Harris, *Hegel's Development*, xvi.
11. Hegel to Niethammer, February 12, 1809, and October 23, 1812, *Letters*, 191; quote in Harris, *Hegel's Development*, 3; *Letters*, ed.'s intro., 34; 530, 364–365, 476–467, 475–516 passim; Kaufmann, 40, 47, 478.
12. Karl Löwith, *From Hegel to Nietzsche: The Revolution in Nineteenth-Century Thought*, trans. David E. Green (New York: Rinehart and Winston, 1964). A basic question of this work is Hegel's putative Christianity. Also see Fackenheim, 155.
13. *Philosophy of Religion*, 3:63–64; *History of Philosophy*, 3:80; *Philosophy of Religion*, 1:126; "The Spirit of Christianity," in *E.T.W.*, 254, 271, 254, 271, 292, on miracles; *History of Philosophy* 3:332, 319; "Positivity of the Christian Religion," in *E.T.W.*, 163, 78-81.
14. *History of Philosophy*, 3:188–216, 191, 208, 216, 194, 211. On self-division of God see *Philosophy of Religion*, 3:292–293.
15. Hegel's review of "[Karl Wilhelm] Solger's *Posthumous Writings and Correspondence*," in *Encyclopedia*, 294.
16. *Philosophy of Religion*, 3:122, 143; *Encyclopedia*, intro., arts. 20, 21; Hegel to [Karl Sigmund] von Altenstein, April 16, 1822, in *Letters*, 392: "Both of these—classical institutions and religious truth—insofar as it [*sic*] still consists in the old dogmatic teachings of the church—I would view as the substantial side of preparation for philosophic study."
17. See Kennedy, chap. 2; cf. Hegel, *Philosophy of Religion*, 1:155–173. The sections devoted to British and French empiricism in the *History of Philosophy*, e.g., 3:323, 332, are very critical of it, as are arts. 32–33 of the *Encyclopedia*.
18. *Phenomenology*, 461: "God is attainable in pure speculative knowledge alone and *is* only in that knowledge, and is only that knowledge itself, for He is Spirit." This sentence simultaneously exalts speculative knowledge and denies Gods ontological existence by the phrase "is only that knowledge itself," which is repeated for emphasis. Besides, Alexandre Kojève, Ernst Bloch, and Thomas Alltizer are cited by Butler and Seiler as interpreting Hegel as an atheist; see *Letters*, intro., 39.
19. *Philosophy of Religion*, 3:65–67, 111, 114; *History of Philosophy*, 3:61–67, 452ff.
20. *Encyclopedia*, 68, 70; *Philosophy of Religion*, 1:179; *Phenomenology*, 292.
21. *Philosophy of Religion*, 3:111, 92; cf. ibid., 144: History "is God's objectification of himself." See *Phenomenology*, 413, on "moments of spirit." Kelly

opines, "And yet only Christianity and the mystery of the Incarnation have made it possible to think the principle of philosophical history at all" (302).
22. *Letters*, 13.
23. *Aesthetics*, 2:822; *History of Philosophy*, 3:408; *Philosophy of Religion*, 1:208; *History of Philosophy*, 3:416; *Phenomenology*, 398.
24. *Philosophy of Religion*, 1:240; 3:109. Both these statements are from manuscripts dating approximately from 1821; the first, in vol. 1, is not dated by the editors. Wherever possible, I have cited the MS versions, as these were written by Hegel as opposed to the *Lectures*, which were recorded by students.
25. *Philosophy of Religion*, 3:115, 144–145, 66.
26. *Philosophy of Religion*, 1:74–75, 77, ed.'s intro.; ibid., 3:132. The preceding reference contains a succinct dialectical rendition of the Redemption: "In its development this [process is] the going forth of the divine idea into the uttermost cleavage, even to the opposite pole of the anguish of death, which is itself the absolute reversal, the highest love, containing the negation of the negative within itself [and being in this way] the absolute reconciliation, the sublation of the prior antithesis between humanity and God. The end is [presented] as a resolution into glory, the festive assumption of humanity in the divine idea." The redemption, while not perhaps an ontological or historical reality for Hegel, is as indelible to human consciousness as are the categories of understanding for Kant. Both have the similitude of being products of the mind rather than the objective world.
27. Philipp Konrad Marheineke's *Basic Teachings of Christian Dogmatics* (2nd ed., 1827) paraphrased in *Letters*, 512; *Philosophy of Religion*, 3:109ff.
28. On gnosticism see Hans Ur van Balthasar's intro. to St. Irenaeus, *The Scandal of the Incarnation*, trans. John Saward (San Francisco: Ignatius Press,1997), 1–11. On Marx, see Chapter 11.
29. Kant, *Critique of Pure Reason* [1787], 2nd ed., trans. Norman Kemp Smith (New York: St. Martin's Press, 1965; orig. publ. 1929), 221–233, 396–421, leaves a place for faith. The infinitive *aufheben* can be variously translated as "encompass," "supersede," "elevate," "cancel out and lift up," "sublimate," "transcend," and "sublate." The advantage of the last two is that they connote inclusion of the canceled elements of the antithesis. Scholars are in general agreement that Hegel did not use the term "synthesis" much at all, especially in this context. The historical school described here, and which is distinct from Hegel's, espoused what became known as "historicism." Hegel resembles it in that for him truth is produced by history. He differs from it in that for him there is an absolute at the end of history, and therefore truth for him exists and has an objective *telos* and referent and is therefore not *simply* "the product of the times," but the product of all time past and future.
30. *Philosophy of Religion*, 1:130; review of Solger appended to *Encyclopedia*, 294; *Aesthetics*, 104.
31. *Letters*, 537–538; Fackenheim, 184; *Philosophy of Religion*, 1:58, 153.
32. "Positivity of the Christian Religion," in *E.T.W.*, 69, 178–179 and passim, 205–220; *Philosophy of Religion*, 3:117; cf. 2:669–681.

33. "Berne Fragment" (1793–1794), in *Three Essays*, 59–64, 74n, 88, 95. The "positivity" of the Christian religion was "something given," namely by an authority, as opposed to the inner religion discovered by oneself and in oneself. *E.T.W.*, 144, 165.
34. "Life of Jesus," in *Three Essays*, 104–165.
35. "Berne Fragment," in *Three Essays*, 83.
36. While he devotes five pages to the seventeenth-century Cartesian priest Nicolas de Malebranche, he is still scornful. *History of Philosophy*, 3:290–295, 94–95, 67; ibid., 61–68, on Abelard. It is possible that Hegel appreciated the audacity of the twelfth-century Abelard and Anselm over the magisterial synthesis of St. Thomas, because Abelard practiced dialectics and Anselm was something of an idealist with his ontological proof of the existence of God. Ibid., 61–69; *Letters*, 95, 51, containing jibes about medieval "forgeries" and describing Jesuits as "vermin."
37. *History of Philosophy*, 3:55–62, 99, 101–102, 151–152, 158–159; *Philosophy of History*, 414.
38. Wahl, 21–47.
39. *History of Philosophy*, 3:14, 148.
40. *Letters*, 327–328. The term "transfer of sacrality" is used by Mona Ozouf, *La fête révolutionnaire (1789–1799)* (Paris: Gallimard, 1976).
41. "Tübingen Essay," in *Three Essays*, 44. Outside France the term "Jacobin" did not necessarily mean a sympathizer with Robespierre or the Terror, but simply one generally favorable to the French Revolution.
42. Fackenheim, 163.
43. *Philosophy of History*, 335.
44. *Phenomenology*, 332–334, 411; Wahl, 53.
45. *Phenomenology*, 96. For a positive evaluation of the Enlightenment, see *Philosophy of History*, 392–397; Fackenheim, 64; *Phenomenology*, 355–364; *Philosophy of History*, 256.
46. "The Positivity of the Christian Religion," in *E.T.W.*, 161.
47. Klaus Epstein, *The Genesis of German Conservatism* (Princeton: Princeton University Press, 1966), chaps. 9, 12, 14; *Letters*, 97–109, 295–308, 358, 377, 444–450, 461–462, 468, 518, on Hegel's moderate but engaged attitude toward the *burschenschaften*, the liberal student associations; Hegel, "The Relationship of Religion to the State according to the Lectures of 1831," *Philosophy of Religion*, 1:451. Hegel's definition of freedom here collapses the distinction between *ought* and *is* just as his moral theory did.
48. *Philosophy of Religion*, 3:373–374. Cf. *Philosophy of History*, 166: "The state is the divine will."
49. *Philosophy of Right*, 279n to paragraph 258; "Positivity of the Christian Religion," in *E.T.W.*, 112; *Philosophy of Right*, 168–169, 172. Just as one could speak of the secularization of the church institutions in the Middle Ages—the use of feudalism to define relations between Rome and vassal states, and later the use of Roman law, the church bureaucracy and armies to defend the papal states—Hegel makes note of the absorption of ethical and religious responsibilities by the state.
50. *Philosophy of Religion*, 3:125. According to his editor, Hegel is here quoting from the second stanza of the "passion hymn" "Taurigkeit O Herzeleid" by

Johannes Rist (1641), which clearly refers to Christ, whose ultimate love for mankind is his death.

## Chapter 11

1. This chapter is used by permission of Blackwell Publishing. The standard English edition of Marx's works used is Karl Marx and Frederick Engels, *Collected Works*, 44 vols. (New York: International Publishers, 1975), hereafter *C.W.*; Solomon F. Bloom, "Karl Marx and the Jews," *Jewish Social Studies* 4 (1942): 3–16; Julius Carlebach, *Karl Marx and the Radical Critique of Judaism* (London: Routledge/Kegan Paul, 1978)—probably the best researched and most judicious interpretation; see also Carlebach, "Karl Marx and the Jews of Jerusalem," *Soviet Jewish Affairs* 2 (1972): 71–74; Joseph Clark, " Marx and the Jews: Another View," *Dissent* 28 (1981):74–86; Sander Gilman, "Karl Marx and the Secret Language of Jews," *Modern Judaism*, 4 (1984): 275–294; Dennis K. Fischman, *Political Discourse in Exile: Karl Marx and the Jewish Question* (Amherst, MA: University of Massachusetts Press, 1991); David B. Ingram, "Rights and Privileges: Marx and the Jewish Question," *Studies in Soviet Thought* 35 (1988): 125–145; Henry Patcher, "Marx and the Jews," *Dissent* 26 (1979): 450–467; Yoav Peled, "From Theology to Sociology: Bruno Bauer and Karl Marx on the Question of Jewish Emanciaption," *History of Political Thought* 13 (1992): 463–485; Enzo Traverso, *The Marxists and the Jewish Question: The History of a Debate (1843–1943)*, trans. Bernard Gibbons (Atlantic Highlands, NJ: Humanities Press, 1994; orig. publ. 1990); Robert S. Wistrich, "Karl Marx and the Jewish Question," *Soviet Jewish Affairs* 4 (1974): 53–60. For general treatment see Jacob Katz, *Out of the Ghetto: The Social Background of Jewish Emancipation, 1770–1870* (Cambridge: Harvard University Press, 1978); Arthur Hertzberg, *The French Enlightenment and the Jews* (New York: Columbia University Press, 1968).
2. Henri Grégoire, *Essai sur la régénération physique, morale et politique des Juifs* (Metz, 1789; Paris: Editions d'Histoire Sociale, 1968), 188–189.
3. B. Blumenkrantz and A. Soboul, "La Révolution française et les juifs: Avant propos," *Annales historiques de la Révolution française*, no. 223 (1976): 10.
4. F. Delpech, "L'Histoire des Juifs en France de 1780 à 1840," *Annales historiques de la Révolution française*, no. 223 (1976): 10ff.
5. Timothy Tackett, *Religion, Revolution, and Regional Culture in Eighteenth-Century France: The Ecclesiastical Oath of 1791* (Princeton: Princeton University Press, 1986), 210–218.
6. Delpech, 10–12. The Terror of 1793/94 had been harsh on Jews, as on all religionists. Synagogues were closed, Sabbath observance was prohibited, and the property of wealthy Jews in Alsace was confiscated.
7. Ibid., 16–19.
8. Werner E. Mosse, "From 'Schutzjuden' to 'Deutsche Staatsbürger Jüdischen Glaubens': The Long and Bumpy Road of Jewish Emancipation in Germany," in *Paths of Emancipation: Jews, States and Citizenship*, ed. Pierre Birnbaum and Ira Katznelson (Princeton: Princeton University Press, 1995), 60–65; Edmund Silberner, "Charles Fourier on the Jewish Question," *Jewish Social Studies* 8(1946): 245–266.

9. Katz, 56, 46; Jonathan Frankel, "Assimilation and the Jews in Nineteenth-Century Europe: Towards a New Historiography?" in *Assimilation and Community: The Jews in Nineteenth-Century Europe*, ed. Jonathan Frankel and Steven J. Zipperstein (Cambridge: Cambridge University Press, 1992), 1–37 passim; Arnulf Zweig, ed., trans., *Correspondence of Kant* (New York: Cambridge University Press, 1999), nos. 12, 13, 22, 49, 52, 65.
10. Katz, 48–64.
11. Quoted in Reinhard Rürup, "The Tortuous and Thorny Path to Legal Equality: 'Jew Laws' and Emancipatory Legislation in Germany from the Late Eighteenth Century," *Leo Baeck Institute Yearbook*, 31 (1986):, 10, 14; Salo W. Baron, "The Impact of the Revolution of 1848 on Jewish Emancipation," in *Emancipation and Counter-Emancipation*, ed. Abraham G. Duker and Meir Ben-Horin (New York: Ktav Publishing House, 1974), 148–194; Marc Crapez, *L'Antisémitisme de gauche au XIXe siècle* (Paris: Berg Institute, 2002), 26, 41. According to Paul R. Sweet, "Fichte and the Jews: A Case of Tension between Civil Rights and Human Rights," *German Studies Review* 16 (1993): 37–48.
12. Jerrold Seigel, *Marx's Fate: The Shape of a Life* (Princeton: Princeton University Press, 1978), chap. 2. This is a superior intellectual biography of Marx. See also Lawrence S. Stepelevich, ed., *The Young Hegelians: An Anthology* (Atlantic Highlands, NJ: Humanities Press, 1997; orig. publ. 1983), 21–51.
13. The classic on the subject is Karl Löwith's *From Hegel to Nietzsche: The Revolution in Nineteenth-Century Thought* (New York: Holt, Rinehart and Winston, 1964). See also Stepelevich's *Young Hegelians* and David McLellan's pithy and lucid account, *The Young Hegelians and Karl Marx* (London: Macmillan, 1969). Kant's and Hegel's anti-Semitism has been the subject of recent study. See Michael Mack, *German Idealism and the Jew: The Inner Anti-Semitism of Philosophy and German Jewish Responses* (Chicago: University of Chicago Press, 2003).
14. Seigel, chap. 3; Robert C. Tucker, ed., *The Marx-Engels Reader*, 2nd ed. (New York: W. W. Norton, 1978), 9–12; Marx, *Critique of Hegel's Philosophy of Right*, trans. and ed. Joseph O'Malley (Cambridge: Cambridge University Press, 1978; orig. publ. 1970). This critique of Hegel's idealism is continued by Marx and Engels in the *German Ideology* of 1845.
15. Feuerbach, *The Essence of Christianity*, trans. George Eliot (New York: Harper and Row, 1957), 338–339, 226. For Marheineke's quote see his *Basic Teachings of Christian Dogmatics* (2nd ed., 1827) paraphrased in Hegel, *The Letters*, trans. Clark Butler and Christiane Seiler, ed. Clark Butler (Bloomington, IN: Indiana University Press, 1984), 512.
16. Feuerbach, 160, 153, 155.
17. Stepelevich, 35, 30. For an intensive study of Bauer's intellectual influence on Marx, which the author contends was greater than that of Feuerbach, see, Zvi Rosen, *Bruno Bauer and Karl Marx: The Influence of Bruno Bauer on Marx's Thought* (The Hague: Martinus Nijhoff, 1977), 73, 75, 107, 118, 124, 136, 140, 138, 148–149, 156, 159, 173, 199, 222.
18. Robert M. Bigler, *The Politics of German Protestantism: The Rise of the Protestant Church Elite in Prussia, 1815–1848* (Berkeley: University of California Press, 1972), 88, 107, 124, 127, 148, 149; John E. Groh, *Nineteenth Century German Protestantism: The Church as Social Model* (Washington, DC: University Press of

258 / NOTES

America, 1982), 200, 76–77, 211–212; Nicholas Hope, *German and Scandinavian Protestantism, 1700–1918* (Oxford: Clarendon Press, 1995), 413–414, 419, 489; Peled, 465. On Bauer see Richard Leopold, "The Hegelian Anti-Semitism of Bruno Bauer," *History of European Ideas* 25 (1999): 179–206. See also C. M. Clark, *The Politics of Conversion: Missionary Protestantism and the Jews in Prussia, 1728–1941* (Oxford: Oxford University Press, 1995).

19. Marx, *On the Jewish Question*, *C.W.* 3:158 (hereafter *J.Q.*). Marx's essay, of course, was a response to Bruno Bauer's *Die Judenfrage* (Brunswick, 1843).
20. "Comments on Latest Prussian Censorship Instruction," *C.W.* 1:116–119.
21. Marx, *C.W.* 1:199.
22. Ibid., 1:118, 201, 188, 274–275.
23. *J.Q.*, 150–151, 160–161. All the emphases in this chapter are Marx's.
24. Stepelevich, 187–197, 197.
25. Bruno Bauer, "The Capacity of Present-Day Jews and Christians to Become Free," trans. Michael P. Malloy, *Philosophical Forum* 8, nos. 2–4 (Boston, n.d.), 138, 140–141.
26. *J.Q.*, 147.
27. Seigel, 100–103, 279–289; Marx to Engels, May 10, 1861, and July 30, 1862, in Saul K. Padover, ed., trans., *The Letters of Karl Marx* (Englewood Cliffs, NJ: Prentice-Hall, 1979), 459, 466–467. *J.Q.*, 174.
28. *J.Q.*, 150–156, 152 and passim.
29. *J.Q.*, 151, 156.
30. *J.Q.*, 156, 158. On page 152 he describes the divided condition of man with regard to religion and the state in the United States: "Even if he proclaims himself an atheist through the medium of the state, that is if he proclaims the state to be atheist, [he] still remains in the grip of religion," which subsists on the civil level. Marx is not satisfied with the *separation* of religion and the state but wants a total separation of religion from humanity—in other words, the abolition of religion.
31. The words "return of man into himself" are found most noticeably in Marx's *Economic and Philosophic Manuscripts of 1844* (Moscow: Progress Publishers, 1959; orig. publ. 1932). On the relation of Marx and Marxism to national and ethnic questions in the twentieth century, see Frank E. Manuel, "A Requiem for Karl Marx," *Daedalus* 121 (1992): 1–19. Warren Breckman, *Marx, The Young Hegelians, and the Origin of Radical Social Theory* (Cambridge: Cambridge University Press, 1999), 299–308, shows the hostility of the Young Hegelians to Hegel's Christian personalism and personal conception of sovereignty, of which, we can add, Marx's "species-being" is the antithesis.
32. *J.Q.*, 159.
33. Ibid., 155–156.
34. Bauer, "The Jewish Problem," in Stepelevich, 188; *J.Q.*, 149, 159; Manuel, 1–19.
35. *J.Q.*, 168.
36. Ibid., 173.
37. Rosen, 225ff.
38. Rosen, 140, 180, 195, 202–294, 212, 91. Bauer's and Marx's use of inversion is called the "transformative method." This is a characteristic "sleight of hand" among German idealists; *J.Q.*, 168–169.
39. Rosen, 216, 222.

40. *J.Q.*, 172, 168–174. The mostly economic anti-Semitism is documented archivally by Eleonore Sterling, *Er ist wie Du: Aus der Frühgeschichte des Antisemitismus in Deutschland (1815–1850)* (Munich: Chr. Kaiser Verlag, 1952). Sterling also documents traditional Christian anti-Semititic accusations that were then current, such as Christ killing (129–143).
41. Ibid., 168–169; Marx and Engels, *The Holy Family, or Critique of Critical Criticism against Bruno Bauer and Company* [1844], in *C.W.* 4:89, 88–115 passim, where they elaborate on themes first enunciated in *On the Jewish Question*. Shlomo Avineri, "Marx and Jewish Emancipation," *Journal of the History of Ideas* 25 (1964): 449.
42. Marx and Engels, *The Holy Family*, *C.W.* 4:108, 110, 112. Friedrich Herr, *God's First Love: Christians and Jews over Two Thousand Years* (New York: Waynwright and Talley, 1967), 194, has some interesting words on the alien nature of Mammon to traditional Judaism.
43. *The Holy Family*, *C.W.* 4:92–95.
44. *J.Q.*, 174.
45. Ibid.
46. See the informative article by Alan Mittleman, "From Jewish Street to Public Square," *First Things*, no. 125 (August 2002): 29–37.

## Chapter 12

1. Sources used include the following: Nikolai Chernyshevsky, *What Is to Be Done?* trans Michael R. Katz, ed. William A. Wagner (Ithaca, NY: Cornell University Press, 1989); Fyodor Dostoyevsky, *Complete Letters*, trans. and ed. David A. Lowe, 5 vols. (Ann Arbor: Addis, 1991), hereafter *Letters*, followed by the volume number (in Roman) and the letter number (in Arabic); *Diary of a Writer*, trans. and ed. Boris Brasol (New York: Braziller, 1954; orig. publ. 1949), hereafter *Diary*; *The Notebooks for The Possessed*, ed. and intro. Edward Wasiolek, trans. Victor Terras (Chicago: University of Chicago Press, 1968), hereafter *Notebooks*; *Notes from Underground*, trans. and ed. Michael R. Katz (New York: W. W. Norton, 1989); *Demons*, trans. Richard Pevear and Larissa Volokhonsky, intro. Joseph Frank (New York: Knopf, 2000) original pub. 1994, *Demons* is a new translation of *The Possessed*; Ivan Turgenev, *Fathers and Sons*, trans. Constance Garnett (New York: Airmont, 1967); Robert Belknap, "Shakespeare and *The Possessed*," *Dostoevsky Studies* 5 (1984): 63–69; James H. Billington, *Fire in the Minds of Men: Origins of the Revolutionary Faith* (New York: Basic Books, 1980); Jacques Catteau, "Le Christ dans le mirroir des Grotesque ('Les Demons')," *Dostoevsky Studies* 4 (1983): 29–36; Denis Dirscherl, S. J., *Dostoievsky and the Catholic Church* (Chicago: Loyola University Press, 1986); Sigurd Fasting, "Dostoievskij and George Sand," *Russian Literature* 4 (1976), 309–321; Joseph Frank, *Dostoievsky*, vol. 1, *The Seeds of Revolt, 1821–1849* (Princeton: Princeton University Press, 1976); Frank, *Dostoievsky*, vol. 2, *The Years of Ordeal, 1850–1859* (Princeton: Princeton University Press, 1983); Frank, *Dostoievsky*, vol. 3, *The Stir of Liberation, 1860–1865,* (Princeton: Princeton University Press, 1986); Frank, *Dostoievsky*, vol. 4, *The Miraculous Years, 1865–1871* (Princeton: Princeton University Press, 1995). The last volume of this superb

biography is titled *Dostoievsky*, vol. 5, *The Mantle of the Prophet, 1871–1881* (Princeton: Princeton University Press, 2002). Ina Fuchs, *Die Herausforderung des Nihilismus: Philosophische Analysen zu F. M. Dostojewskijs Werk "Die Dämonen"* (Munich: O. Sagner, 1987); David I. Goldstein, *Dostoyevsky and the Jews* (Austin: University of Texas Press, 1981; orig. French publ. 1976); Serge V. Gregory, "Dostoevsky's *The Devils* and the Antinihilist Novel," *Slavic Review* 38 (1979): 444–455; Michael Holquist, *Dostoyevsky and the Novel* (Princeton: Princeton University Press, 1977); Vyacheslav Ivanov, *Freedom and the Tragic Life: A Study in Dostoyevsky*, foreword Sir Maurice Bowra (New York: Noonday Press, 1952); Malcolm V. Jones, "Dostoievsky and Europe: Travels in the Mind," *Renaissance and Modern Studies* 24 (1980): 38–57; Jones, *Dostoyevsky after Bakhtin: Readings in Dostoyevsky's Fantastic Realism* (Cambridge: Cambridge University Press, 1990); Jones, "Some Echoes of Hegel in Dostoyevsky," *Slavonic and East European Review* 49 (1971): 500–520; Geir Kjetsaa, "Dostoevsky and His New Testament," *Dostoevsky Studies* 4 (1983): 95–112; Liza Knapp, *The Annihilation of Inertia: Dostoevsky and Metaphysics* (Evanston, IL: Northwestern University Press, 1996); Marine Kostalevsky, *Dostoyevsky and Soloviev: The Art and Integral Vision* (New Haven, CT: Yale University Press, 1997); Janko Lavrin, "A Note on Nietzsche and Dostoevsky," *Russian Review* 28 (1969): 160–170; W. J. Leatherbarrow, "Apocalyptic Imagery in *The Idiot* and *The Devils*," *Dostoevsky Studies* 3 (1982): 43–51; Leatherbarrow, *Fyodor Dostoevsky: A Reference Guide* (Boston: G. K. Hall, 1990); Waclaw Lednicki, "Europe in Dostoevsky's Ideological Novel," in *Russia, Poland and the West: Essays in Literary and Cultural History* (London: Hutchinson, 1954), 133–179; Gordon Livermore, "Stepan Verkhovensky and the Shaping Dialectic of Dostoevsky's *Devils*," in *Dostoevsky: New Perspectives*, ed. Robert Louis Jackson (Englewood Cliffs, NJ: Prentice Hall, 1984); Ralph E. Matlaw, "Chronicler of *The Possessed*," *Dostoevsky Studies* 5 (1984): 37–47; Robin Feuer Miller, "Imitations of Rousseau in *The Possessed*," *Dostoevsky Studies* 5 (1984): 77–89; Gary Saul Morson, "What Is the Intelligentsia? Once More, an Old Russian Question," *Academic Questions* 6 (1993): 20–38; Gene M. Moore, "The Voices of Legion: The Narrator of *The Possessed*," *Dostoevsky Studies* 6 (1985): 51–65; Charles Moser, *Antinihilism in the Russian Novel of the 1860s* (The Hague: Mouton, 1964); Harriet Murav, *Holy Foolishness: Dostoyevsky's Novels and the Poetics of Cultural Critique* (Stanford: Stanford University Press, 1992); Nadine Natov, "The Theme of Chantage (Blackmail) in *The Possessed*: Art and Reality," *Dostoevsky Studies* 6 (1985): 3–33; Joyce Carol Oates, "Tragic Rites in Dostoyevsky's *The Possessed*," in *Contraries: Essays* (New York: Oxford University Press, 1981), 17–50; Derek Offord, "Dostoyevsky and Chernyshevsky," *Slavonic and East European Review* 57 (1979): 509–530; Konrad Onasch, "Dostojevskij in der Tradition de russischen, Laientheologen," *Dostoevsky Studies* 4 (1983): 113–124; Onasch, "F. M. Dostoevskij: Biographie und Religiöse Identität: Versuch einer Synpose," *International Journal of Slavic Linguistics and Poetics* 31–32 (1985): 295–308; Onasch, "Die 'Gerechten' und 'Stillen im Lande': Zur Kirchenkritik des 19. Jahrhunderts bei Leskov und Dostoevskij," *Dostoevsky Studies* 8 (1987), 135–142; Rado Pibric, "Notes from the Underground One Hundred Years after the Author's Death," in *Dostoievski and the Human Condition after a*

*Century*, ed. Alexj Ugrinsky, Frank S. Lambasa, and Valija K. Ozolins (Westport, CT: Greenwood Press, 1986), 71–77; Philip Rahv, "The Other Dostoyevsky," in *Essays in Literature and Politics, 1932–1972*, ed. Arabel J. Porter and Andrew J. Dvosin (Boston: Houghton Mifflin, 1978), 186–207; Rahv, "Dostoyevsky in *The Possessed*," ibid., 107–128; Dennis Patrick Slattery, "Idols and Icons: Cosmic Transformation in Dostoevsky's *The Possessed*," *Dostoevsky Studies* 6 (1985): 35–50; Franco Venturi, *Roots of Revolution: A History of the Populist and Socialist Movements in Nineteenth-Century Russia* (Chicago: University of Chicago Press, 1983; orig. publ. 1960); René Wellek, "Bakhtin's View of Dostoevsky: 'Polyphony' and 'Carnivalesque,'" *Dostoevsky Studies* 1 (1980): 163ff; Bruce K. Ward, *Dostoyevsky's Critique of the West: The Quest for the Earthly Paradise* (Waterloo, ON: Wilfrid Laurier Press, 1986); Edmund Wilson, "Dostoevsky Abroad," in *Shores of Light: A Literary Chronicle of the Twenties and Thirties* (New York: Farrar, Straus and Young, 1952), 408–414.
2. *Notes from Underground*, passim, for his early disorientation.
3. *Letters*, V, 731, 742.
4. *Diary*, 779–781; Frank 1:29, 50; vol. 2, chap. 9.
5. Frank, 1:50.
6. *Diary*, 209–210; Frank, 2:89, 144.
7. *Diary*, 906.
8. *Letters*, III, 360.
9. *Diary*, 651; *Letters*, I, 13.
10. *Diary*, 911; Lednicki, 159.
11. Jones, "Dostoyevsky and Europe," 44; *Diary*, 371. It is well known that Dostoyevsky had epilepsy. He suffered a number of attacks during his European sojourn of 1867–1871. According to Frank, he could reach a state of complete calm and reconciliation with his world during certain stages of these attacks.
12. Jones, "Some Echoes of Hegel"; Venturi, 16; Feuer Miller, 79.
13. Lednicki, 164, 167; Frank, 2:160.
14. Quoted in Frank, 1:263. On Dostoyevsky's intellectual similitude to Nietzsche on certain points, see Lavrin, 160–170.
15. Venturi, 79, 82, 87, 89, 96, 99, 108; Frank, 1:149; 2:232; 1:257–269.
16. Turgenev, 92, 114, 124, 161, 163; Flaubert, *Madame Bovary* (New York: Charles Scribner's Sons, 1958; orig. publ. 1930), II, xi.
17. Turgenev, 190.
18. Chernyshevsky, 17, 20, passim; Billington, 393.
19. Offord, 509–530; Leatherbarrow, "Apocalyptic Imagery," 45.
20. Venturi, 380–381; Billington, 389–400, 546n160; Venturi, 365–366. Dostoyevsky was obsessed by what he considered to be the necessary link between atheism and criminality (and in *Demons*, nihilism). In his last novel, *The Brothers Karamazov* (1879/80), Ivan says that without the immortality of the soul, any action is licit "for every individual, like ourselves, who does not believe in God or immortality, the moral law of nature must immediately be changed into the exact contrary of the former religious law, and that egoism, even to crime, must become, not only lawful but even recognised as the inevitable, the most rational, even honourable outcome of his position."

"Everything would be lawful," Dostoyevsky pleads, "even cannabilism." *The Brothers Karamazov*, trans. Constance Garnett (New York: Random House, Modern Library, n.d.), bk. II, chap. 6, 69; cf. bk. III, chap. 8, 138, and bk. V, chap. 3, 243. See also David Cortesi, "Dostoyevsky Didn't Say It: Exploring a Widely-Propagated Misattribution ["If God does not exist, everything is permitted"], http://www.infidels.org/library/modern/features/2000/Cortesi1.html. Cortesi is technically correct. Dostoyevesky did not say those last words exactly. But Cortesi is essentially wrong and misleading in that Dostoyevsky said basically the same thing in the words quoted above in the Constance Garnett translation. I am grateful to my former student Thomas Kaiser and to the Russian literature scholar Aleksandr Klimoff for help in tracking down this quotation.

21. Bayle, *Historical and Critical Dictionary*, ed. Richard Popkin (Indianapolis, IN: Hackett, 1991), 399–408; Dostoyevsky, *Notebooks*, 85.
22. Knapp, 118; *Notebooks*, 226.
23. *Notebooks*, 179, 173, 209, 241; Wellek, 163ff.
24. Kjetsaa, 100. We are not alone in thinking that secularism is on Dostoyevsky's mind; see ibid., Dostoyevsky "realized that the Arian heresy, that is to say the denial of the divinity of Christ, was the first step in the European process of secularization" which was his calling to oppose. In the West this would have been deism. Also, Bruce K. Ward in *Dostoyevsky's Critique* elucidates the utopianism of the early nineteenth century in the West, where "world citizenship and universal love were preached by its intellectuals such as H. Saint-Simon as a 'New Christianity'" (27, 50, 47). Liberalism was being challenged since mid-century on the one hand by the nihilism of Nechaev and *the* self-deification of the God-man Kirillov in *Demons*, and on the other hand by the pope, represented by *The Brothers Karamazov*'s Grand Inquisitor, the guarantor of bread and security minus freedom, i.e., communism (97, 99). "Nechaev's crime was in [Dostoyevsky's] eyes yet another sign that Russia had entered the apocalyptic age of socialism" (Kjetsaa, 100, 104). For the sources of Dostoyevsky's religious and apocalyptic thinking see Onasch, 113–124, and Leatherbarrow, "Apocalyptic Imagery," 43–51.
25. *Notebooks*, 251, 417; Knapp, 107.
26. *Demons* (hereafter *D.*), 7–10.
27. Venturi, 24; *D.*, 56; Frank, 4:467, 454–455; Dostoyevsky, *Notes from Underground*, 12.
28. *D.*, 15, 19, 300.
29. *D.*, 24, 27.
30. *Letters*, III, 387; Frank, vol. 1, chap. 13; Livermore, 177; Frank, 1:255; 4:465, 486; *Notes from Underground*, 89.
31. *D.*, 58, 304.
32. *D.*, 455.
33. *D.*, 485.
34. *D.*, 516.
35. *D.*, 513; Catteau, 30; Slattery, 35–50; Gregory, 452; Wellek, 163ff; Leatherbarrow, "Apocalyptic Imagery," 43–51.
36. *D.*, 407.
37. *D.*, 366, 370–371.

38. *D.*, 313, 462, 515.
39. Moser, chaps. 2 and 4.
40. Frank, 3:172; *D.*, 418, 389, 508, 515.
41. Dirschel, 93; *D.*, pt. 3, chap. 4.
42. *D.*, pt. 1, chap. 2; pt. 3, chap. 3.
43. *D.*, 371, 402; Rahv, 117; Morson, 25; Rahv 112 states that "the example of Dostoyevsky may be put in the struggle against secular ideas" (112).
44. *D.*, 688, 701, 709, 711, 713.
45. *D.*, 405, 554; Billington, 546.
46. *D.*, 581–582.
47. For a comparison of Nietzsche's and Dostoyevsky's nihilism, see Fuchs, esp. 244–265. A similar comparison could be made of the two thinkers' views of socialism as a secularization of Christianity.
48. *D.*, 663.
49. *D.*, 678.
50. *Letters*, I, 5, 7, 431–432, 494–495.
51. There were contemporary Christian socialists in the nineteenth century such as Philippe Buchez (1796–1865) and Félicité de Lamennais (1782–1854). One cannot but help thinking, however, that the socialist view of human welfare as largely material, of culture as wholly anthropocentric, and of the universe as wholly immanent is certainly quite secular and often atheistic. On the religion of man, see the author's "The French Revolution and the Genesis of a Religion of Man, 1760–1885," in *Modernity and Religion*, ed. Ralph McInerny (Notre Dame, IN: University of Notre Dame Press), 61–88. See n. 25 and the next chapter on Solzhenitsyn.

## Chapter 13

1. Solzhenitsyn's works used include the following: *East and West*, trans. Janis Sapiets (New York: Harper and Row/Perennial Library, 1980); *For the Good of the Cause* (London: Sphere Books, 1971; orig. publ. 1964); *The Gulag Archipelago, 1918–1956: An Experiment in Literary Investigation*, trans. Thomas P. Whitney, vols. 1 and 2, pts. 1–6; Harry Willetts, vol. 3, pts. 6 and 7 (New York: Harper and Row, 1973–1976), hereafter abbreviated as *G.*; *Invisible Allies*, trans. Alexis Klimoff and Michael Nicholson (Washington, DC: Counterpoint, 1995); *A Letter to the Soviet Leaders* (New York: Harper and Row, 1975; orig. publ. 1974); *The Mortal Danger: How Misconceptions about Russia Imperil America*, trans. Michael Nicholson and Alexis Klimoff (New York: Harper and Row, 1980); *Nobel Lecture*, trans. F. D. Reeve (New York: Farrar, Straus and Giroux, 1972); *One Day in the Life of Ivan Denisovich*, trans. Ralph Parker, intro. Marvin L. Kalb, foreword Aleksandr Tvardovsky (New York: Signet / Penguin Books, 1973; orig. publ. 1963); *The Oak and the Calf: A Study of Literary Life in the Soviet Union*, trans. Harry Willetts (New York: Harper and Row, 1980; orig. publ. 1975); *Warning to the West*, trans. Harris L. Coulter and Nataly Martin, ed. Alexis Klimoff (New York: Farrar, Straus and Giroux, 1976; orig. publ. 1975); *From under the Rubble: Two Short Novels*, ed. Michael Scammel, intro. Max Hayward (Boston: Little and Brown, 1975); "*We Never Make Mistakes,*" trans. Paul W. Blackstone (New York: W. W. Norton, 1971; orig.

publ. 1963); *A World Split Apart: Commencement Address Delivered at Harvard University, June 8, 1978* (New York: Harper and Row, 1978) and "Address to the International Academy of Philosophy, September 14, 1973, Liechtenstein," in *The Russian Question at the End of the Century* (New York: Farrar, Straus and Giroux, 1995); "The Big Losers in the Third World War," *New York Times*, June 22, 1974, sec. 4, 15; "Brezhnev Cannot Look a Priest Straight in the Eye," *Christianity Today*, June 6, 1980, 13; "Conversation with Solzhenitsyn," interview by Janis Sapiets, *Encounter*, March 1975, 67–72; "An Interview: Aleksandr Solzhenitsyn," by Janis Sapiets, *Kenyon Review* n.s. 1, no. 3 (Summer 1979): 8–17; "How Things Are Done in the Soviet Provinces," *New York Times*, September 30, 1974, 34; "I am No Russian Ayatollah," *Encounter*, February 1980, 34–35; "Have Men Forgotten God?" *National Review*, July 22, 1983, 872–876; "Solzhenitsyn Speaks Out," *National Review*, June 6, 1975, 603–609; "Aleksandr Solzhenitsyn Talks to Walter Cronkite about his Exile from Russia," *Listener*, July 1974; "Remarks at the Hoover Institution, May 24, 1976," *Russian Review* 36, no. 2 (April 1977): 184–189; "New Letter Explains Call on Soviets to Abandon Aggression and Lay Down Arms," *Congressional Record* 120, pt. 7 (April 1, 1974): 024–026; "Schlesinger and Kissinger," *New York Times*, December 1, 1975, 3; "Solzhenitsyn and the KGB," *Time*, May 27, 1974, 51; "Solzhenitsyn Says Soviet Is 'Serfdom,'" *New York Times*, April 6, 1974, 6; "Men Have Forgotten God" [an adaptation of the address at Buckingham Palace after accepting the Templeton Prize of 1972], *National Review*, July 22, 1983, 872; Solzhenitsyn, *A Lenten Letter to Pimen, Patriarch of All Russia* (Minneapolis: Burgess, 1972). The author wrote this chapter before Solzhenitsyn's *Two Hundred Years Together* appeared in English and before he read Anne Applebaum's *Gulag: A History of the Soviet Camps* (New York: Allen Lane/Penguin, 2003). Based on Russian archival research, it is an indispensable complement to Solzhenitsyn's oral history.
2. D. M. Thomas, *Aleksandr Solzhenitsyn: A Century in His Life* (New York: St. Martin's Press, 1998), 3–51; Michael Scammell, *Solzhenitsyn: A Biography* (New York: W. W. Norton, 1984), chaps. 1–2. Edward Ericson, Jr. and Daniel J. Mahoney's. *The Solzhenitsyn Reader* (ISI) was in press at the time of this publication.
3. Thomas, 41–51, 77, 93–141, 164–173; Scammell, 126–159.
4. E.g., *G.* 2:355.
5. *G.* 1:245–246, 251–253.
6. *G.* 1:493, 396, 593, 539–541; 3:37, 387–388, 390.
7. *G.*, vol. 1, pt. 1, chap. 3, 493, 535–541; 3:387–388, 390; Paul Johnson, *Modern Times: The World from the Twenties to the Nineties* (New York: Harper and Row, 1983), 275–276.
8. *G.* 2:181, 246, 340.
9. *G.* 1:171.
10. According to Stephane Courtois et al., *The Black Book of Communism: Crimes, Terror, Repression* (Cambridge, MA: Harvard University Press, 1999), 234, five thousand women and children languished in camps in 1948.
11. *G.* 1:54–55, 435–439, 324. Solzhenitsyn's figures may be compared to those of Stephane Courtois et al, *Black Book*, 151–167, 190–207, 231, 234, 256, 263: liquidated "kulaks" amounted to 15 million, of which 6 million starved to

death. Nearly 700,000 party members were executed in the purges of the thirties, of which 90,000 were fatally tortured. Seven million died in the camps between 1934 and 1940. In 1953 there were still 1.2 million *zeks* in the camps; 310,00 were freed after the 20th Party Congress in 1956; 600,000 died as deportees or "displaced persons." Solzhenitsyn's total tally way exceeds courtois' of 20 million.
12. *G.* 2:548; 3:28, 413. Nikolai Vasilyevich Krylenko (1885–1938) was the chief state prosecutor from 1918 to 1931 and then people's commissar of justice until he too was shot.
13. "Letter to the Soviet Leaders" [1974], in *East and West*, 92; *G.* 3:459.
14. "Solzhenitsyn Speaks Out," 607. Cf. Solzhenitsyn's "Les Deux Révolutions," *La Lettre Internationale*, Autumn 1989—a detailed comparison of the French revolutions of 1789 and 1793 and the Russian Revolution of 1917 and after. He believes that neither revolution brought liberty, but only the promise or ideology of liberty and equality. See Daniel J. Mahoney's perceptive and informative *Aleksandr Solzhenitsyn: The Ascent from Ideology* (Lanham, MD: Rowman and Littlefield, 2001), 62n23, 154n25. On September 25, 1993, the occasion of the 200th anniversary of the massacres of tens of thousands of peasants of the Vendée region by the forces of the French National Convention, Solzhenitsyn delivered a speech at Lucs-sur-Boulogne in which he sought to diabuse those who thought that "revolution could change human nature for the better" (ibid., 62–63n23). Solzhenitsyn probably helped to spearhead some of the disaffection with the French Revolution that was noticeable in the Bicentennial of 1989. Some preferred to speak of "commemoration" rather than "celebration." See the book devoted to the bicentennial by Steven Lawrence Kaplan, *Farewell Revolution: Disputed Legacies, 1789–1799* (Ithaca, NY: Cornell University Press, 1995).
15. *G.* 1:308; 3:365.
16. *G.* 2:355.
17. "Conversation with Solzhenitsyn," 67, 72.
18. *G.* 1:168.
19. *Ivan Denisovich*, 155. See Thomas, 176–177, 185, 199, 360–361, 381–382, 393–393, on Natalya Reshetovskaya, Solzhenitsyn's first wife.
20. *Oak and Calf*, 37, 42, 88. Vladimir Lakshin's firsthand scathing analysis of Solzhenitsyn and his relationship with Tvardovsky in *Solzhenitsyn, Tvardovsky and Novy Mir* (Cambridge, MA: MIT Press, 1980). Lakshin portrays Solzhenitsyn as vainglorious and self-centered.
21. *Oak and Calf*, 38, 88.
22. Marvin Kalb, "Introduction," *Ivan Denisovich*, ix.
23. *Oak and Calf*, passim; Scammell, *The Solzhenitsyn File* (Chicago: Edition Q, 1995), 1, 7, 17, 13, 34–35, 68, 89, 105, 129, 113.
24. *Oak and Calf*, 84.
25. *Literaturnaya Gazeta*, November 12, 1969, translated in Leopold Labedz, *Solzhenitsyn: A Documentary Record* (New York: Harper and Row, 1971), 218; Solzhenitsyn to the Secretariat of the Soviet Writers' Union, 12 November 1969. I owe to Daniel J. Mahoney this last observation.
26. *Oak and Calf*, 10, passim; *Invisible Allies*, passim.
27. *Invisible Allies*, 124.

28. *Invisible Allies*, 79, 81–83, 136–137; cf. *Oak and the Calf*, 530, 118–120, 124, 530.
29. *Oak and Calf*, 349, 304, 157; *Invisible Allies*, 124.
30. *Oak and Calf*, 144, 145, 151.
31. Ibid., 157.
32. Ibid., 219, 320n.
33. Ibid., 220, 218.
34. Ibid., 219, 220–221, 271.
35. Ibid., 289–234 passim.
36. *Nobel Lecture*, passim.
37. *Oak and Calf*, 290, 333–334.
38. Ibid., 180, 197, 218; G. 3:28.
39. *Lenten Letter to Pimen*, 5, 7. For a brief overview see Niels C. Nielsen, *Solzhenitsyn's Religion* (Nashville: Thomas Nielson, 1975), which includes Solzhenitsyn's "Letter to the Third Council of the Russian Churches Abroad," 142–158.
40. *Oak and Calf*, 439; see above n. 21.
41. *Oak and Calf*, 383–453 passim; Scammell, 838–846.
42. "Letter to the Soviet Leaders," 75–142 passim and 95. This letter, dated September 5, 1973, was first published in 1974. Also see "Conversation with Solzhenitsyn," 67; *Warning to the West*, 108–110.
43. *Letter to the Soviet Leaders*, 100. Piotr Arkadevich Stolypin (1862–1911) was the Russsian premier from 1906 to 1911 and simultaneously fought terrorism and advocated return to the land. He was assassinated.
44. *Letter to the Soviet Leaders*, 139.
45. *Oak and Calf*, 533; *From under the Rubble*, 274.
46. *Warning to the West*, 105. In 2001 Solzhenitsyn published a major work, *Two Hundred Years Together*—a history of Russian-Jewish relations. It has not yet been published in English.
47. *Warning to the West*, 109.
48. Nicholson, 388; Solzhenitsyn [the first of five speeches delivered to the Americans and the British in 1975–1976] in *A Warning to the West*, 3–50, which includes an introduction by George Meany.
49. Senator Jesse Helms, "Honorary Citizenship for Solzhenitsyn," *East Europe* 23, no. 27 (July 1974): 2–6; Scammell, 918–919.
50. These remarks were made in an interview with Michael Charlton on the BBC on March 1, 1976. Cf. Solzhenitsyn's speeches on the BBC on March 24 and 29, 1976, printed in *Warning to the West*, 110, 114, 146.
51. Citations here to the speech itself are to *A World Split Apart* (a bilingual edition), 25, 35, 37.
52. Ibid., 39, 41.
53. Ibid., 49.
54. Ibid., 51, 53, 57.
55. Ibid., 61. These are the last words of the address. Man can only go upward because he has sunk so low.
56. Ronald Berman, ed., *Solzhenitsyn at Harvard: The Addresses, Twelve Early Responses, and Six Later Reflections* (Washington, DC: Ethics and Public Policy Center, 1980), 131, 60. In a BBC interview, Solzhenitsyn opined that

"the intellectual West's sympathy for the Soviet system is also conditioned by the source of their ideological origins: materialism and atheism." *East and West*, 172.
57. Hook's remarks are found in *Solzhenitsyn at Harvard*, 85–97, and Schlesinger's on 63–71.
58. *East and West*, 131, 134–135.
59. Ibid., 132, 138, 141.
60. "Men Have Forgotten God," 872.
61. Ibid., 874–876.
62. Ibid., 875.
63. *Address to the International Academy of Philosophy, Liechtenstein*, 118, 122.
64. Ibid., 114, 117, 123, 125.
65. James H. Billington's *Fire in the Minds of Men: Origins of the Revolutionary Faith* (New York: Basic Books, 1980) is the best account up to 1917. See Mahoney's *Solzhenitsyn*, chap. 5, on repentance.

## Chapter 14

1. For a valuable recent survey, see Charlotte Allen, *The Human Christ: The Search for the Historical Jesus* (New York: Free Press, 1998), chap. 9.
2. *Conversations with Mgr Escrivá de Balaguer* (Dublin: Scepter Books, 1968), 65.

# Index

Abelard, Peter, 23–24
Absolute Spirit, 226
absolutism, political, 93, 94, 101, 110
*Address to the Nobility of the German Nation* (Luther), 75–76
*Against the Academicians* (St. Augustine), 19
Agathocles, 67, 72
agnosticism, 5
agoro, freedom of, 1
Albert the Great, 12
Anabaptists, 79, 81–83
Anglican Church, 93, 95, 101–2, 108
   orthodoxy, 102–5
Anselm of Canterbury, 3
Antichrist, 76, 82, 190, 197
anti-Semitism, 166–67, 170, 177, 180, 218
Apocalypse, 28
Aquinas, Thomas. *See* St. Thomas Aquinas
Arianism, 190
Aristotle, 11, 12, 18, 24–25, 26–29, 34–36, 87
Aron, Raymond, 227
atheism, 5, 7, 104–5, 110, 150–52, 154, 161–62, 189, 192–94
   nihilism and, 197–200
Augustine. *See* St. Augustine
autonomy, 4–5
Averroes, 5, 24, 27, 28, 150
Avicenna, 27, 32
Avignon papacy, 44, 46, 55
Avineri, Shlomo, 177

*Babylonian Captivity of the Church* (Luther), 76–77
Bacon, Francis, 87
Bakhtin, Mikhail, 42
Bakunin, Michael, 187, 188
baptism, 76–77, 81, 96
Baptists, 95, 96
Baron, Hans, 43, 59
Barth, Karl, 229
Bauer, Bruno, 168, 169–79
   Marx and, 175–77
   principle of inversion, 176
Bayle, Pierre, 94, 105, 109
Beaumont, Christophe de, 125
Beauvoir, Simone de, 229
Belinsky, Vissarion, 186, 193
Bentham, Jeremy, 193
Bismarck, Otto von, 167, 187
Bloch, Maurice, 166
Boccaccio, 42, 51, 61
Boehme, Jason, 151–52
Bolshevism, 199
*Book of the Body Politik* (Pizan), 60–61
Borgia, Cesare, 62, 65, 67, 71
Boyle, Robert, 102
Bracciolini, Poggio, 61
*Brothers Karamazov* (Dostoyevsky), 183, 185
Bruni, Leonardo, 61, 62
Buber, Martin, 229
Burckhardt, Jacob, 42
Burrows, William, 3
Butler, Clark, 152

Cajetan, Cardinal, 74
Calvin, John, 77, 78, 80, 81, 86–87
Calvinism, 74, 79
*Cancer Ward* (Solzhenitsyn), 210–11, 212
canon law, 33
catechism, 122
Catherine of Siena, 55
Catholic Church. *See* Roman Catholic Church
celibacy, 74, 76–78
Chabod, Federico, 58
Charles II (King), 94, 100–1
Charles V (Holy Roman Emperor), 81, 84–85
Charron, Pierre, 94
chastity, 77–78
Chaucer, Geoffrey, 42
Chernyshevsky, Nicholas, 183, 187–88, 198–99, 206
Chesterton, G.K., 227
Christianity
  philosophy of, 23–24, 28, 37
  Rome and, 16–17, 18, 19, 20–21
Cicero, 29, 32, 36, 47–48, 51, 52
*City of God, The* (St. Augustine), 11, 15–17, 20, 30, 32, 37, 71, 171, 225
Cochrane, C.N., 18
Collins, Anthony, 108
Communism, 2, 7–8, 167, 176–77, 179, 187, 189
  class conflict and, 207
  Solzhenitsyn and, 206–8, 216, 218, 220, 222–24
*Communist Manifesto* (Marx), 179
Comte, Auguste, 7, 130, 192, 193
*Concerning Philosophy* (Varro), 18
*Confessions* (Rousseau), 114, 118–19, 123
*Confessions* (St. Augustine), 14, 15–16
consent, 101, 103–4, 109–10
Constantine, 20, 46
  *See also* Donation of Constantine

corporatism, 61, 63–64
Council of Trent, 28, 31
Crimean War, 185
*Critique of Practical Reason* (Kant), 5, 135–36, 138
*Critique of Pure Reason* (Kant), 5, 132–33, 134–36, 137–38, 139, 141, 143
Croce, Benedetto, 58
Curia, 25

Dante Alighieri, 2, 39, 41–55, 68, 69, 223
  admiration of Roman Empire, 57, 61, 63
  Beatrice and, 41, 42–44, 50, 53–54
  classicism of, 49
  decretalists and, 49
  Guelf party and, 43, 45
  individualism and, 42, 48, 51
  troubadour tradition and, 39, 43
  use of vernacular, 39, 41, 43
Darwinism, 194, 197
*De regimine principum* (Giles de Rome), 60
Declaration of the Rights of Man and Citizen, 165, 174
Decretalists, 33
*Defensor Pacis* (Marsilius of Padua), 100
deism, 5, 113, 127–28, 129
Delumeau, Jean, 86
*Demons* (Dostoyevsky), 181, 183–84, 185–91, 193–94, 198, 202
  nihilism in, 197–201
  overview of, 194–97
Descartes, René, 3–4, 23, 29, 87, 89, 91, 94, 132–33, 135
determinism, 188
dialectical logic, 24, 26, 28
*Dialogues Concerning Natural Religion* (Hume), 132
*Diary of a Writer* (Dostoyevsky), 186
Diderot, Denis, 114, 115, 119, 120
*Die Judenfrage* (Bauer), 172

INDEX / 271

Diet of Worms, 75, 81, 88
Diocletian, 15
*Discourses on Livy* (Machiavelli), 58, 64, 66, 69–70
*Discourse on the Sciences and the Arts* (Rousseau), 115–17, 125
dissenters, secular, 2
diversity, 4
*Divine Comedy* (Dante), 39, 41, 42, 45–47, 49, 50–55
Vergil in, 47, 49–50, 52–54
divine law, 36–37
dogma, 121, 122–24, 126, 128
Kant on, 132, 134–37
Dohm, Christian Wilhelm von, 164, 167
Donation of Constantine, 48, 49–50
Donatists, 14, 15, 21
Dostoyevsky, Fyodor, 181, 183–202, 204, 218, 222, 226
criticism of secularism, 201–2
peasant culture and, 184–85, 187, 190
Petrashevsky circle and, 183, 187
Russian Orthodoxy and, 183, 186, 190–91, 201
on Slavophilism, 184–86
"double truth," 5
Dostoyevsky, 2, 3, 7
Dreyfus Affair, 167
dualism, 113, 128–29

Eck, John, 74–75, 86
Edwards, John, 108
Eliot, T.S., 41
*Emile* (Rousseau), 115, 118–21, 124–25
Emperor Henry VII, 45, 46
empiricism, 3, 152, 153
Empiricus, Sextus, 86
Engels, Friedrich, 80, 171, 174
Enlightenment, 4–6, 26, 29, 37, 95, 108–9, 111
Hegel and, 147–48, 150–56

Kant and, 131–34, 135–39, 140
Rousseau and, 113, 115–16, 120, 122, 124–27, 129
Solzhenitsyn, 206, 217, 219–20, 222
Epicureans, 19
*Epistola de Tolerantia* (Locke), 97, 101–3, 105, 107
Erasmus, 76, 85–87
Escrivá, Josemaría, 227
*Essay Concerning Human Understanding* (Locke), 94, 101–2, 105, 107
*Essay on Infallibility* (Locke), 97–98
*Essay on the Physical, Moral, and Political Regeneration of the Jews* (Grégoire), 164
"Essay on Toleration" (Locke), 97–98; *See also*; *Epistola de Tolerantia* (Locke)
*Essence of Christianity* (Feuerbach), 168, 173, 192, 193
Estates-General, 63
European Economic Community (EEC), 2
Eusebius, 32
evil, 30, 32, 35–36
Exclusion Crisis, 101
externalism, 84

faith, 24–25, 26–29, 35–37
Anselm, 26
Aquinas, 24–25, 26–29, 35–36, 37
Augustine, 27
Hegel on, 150–51, 152–53, 156–57, 159, 161
Luther, 73, 76, 78, 86–88
*Fathers and Sons* (Turgenev), 183, 187
feudalism, 44, 46, 65
Feuerbach, Ludwig, 7, 154, 168–69, 173, 175, 177, 179
Dostoyevsky and, 186, 192, 193
Fichte, 153, 155
Fideism, 87, 132

Filmer, Robert, 101
*First Circle* (Solzhenitsyn), 210–11, 212
*First Tract on Government* (Locke), 95–96
Fourier, Charles, 166, 187, 193, 201
Frank, Joseph, 185
Frederick II, 25
*Freedom of a Christian* (Luther), 76
freedom, spiritual, 76–77, 84
French Revolution, 2, 6, 129, 130, 163, 164–65, 170, 179, 226
  Hegel and, 147–48, 159–61
*From under the Rubble* (Solzhenitsyn), 217
Fronde, 100
fundamentalism, 8

Gerle, Dom, 165
Ghibellines, 43–44, 45
Gibbon, Edward, 58
Gilbert, Felix, 59
Gilbert, R.H., 25–26
Giles de Rome, 60
Ginzburg, Carlo, 86
glasnost, 211
gnosticism, 20
God, existence of, 5–6
grace, 26, 27, 30, 37
Grebel, Conrad, 81–82
Grégoire, Henri, 164–66
Gregory of Tours, 20
Gregory the Great, 20
Grosseteste, Robert, 24
Guicciardini, Francesco, 59, 64, 70
*Gulag Archipelago* (Solzhenitsyn), 208, 210, 211–12, 213–15
  seizure by Soviet government, 216
gulags, 203–6, 208, 209–17, 227

Hegel, Friedrich, 2, 6–7, 29, 58, 133, 135, 141, 142, 147–62, 226
  on Christianity, 150–52
  on consciousness, 153–54, 156, 158–60, 161
  dialectic, 147–48, 152–55, 156, 159
  on externality of Middle Ages, 157–59
  *freiheitskriege,* 160
  on God's existence, 152–55
  on God's "objectivity," 151, 154
  influence, 168–71, 173, 175–76, 178
  Jewish Christian "positivity," 156–58
  Judaism, 153, 156, 158, 161
  on self-consciousness, 154, 156, 159
  speculative philosophy, 148–50, 152, 156
  subjectivist idealism, 154
  on sublation of freedoms, 160
  theology, 151–54, 156–57
Hegelianism, 168–71, 173, 175–76, 178
  Russian, 186, 192–93
Helms, Jesse, 218
Helvétius, 117, 126
heresy, 14, 16, 21, 74, 84
  Aquinas on, 36
Herzen, Alexander, 186, 191–92
*History of Florence* (Machiavelli), 62, 65, 68
*History of Skepticism* (Popkin), 86
Hitler, Adolf, 203, 205, 206, 214, 219, 227
Hobbes, Thomas, 95, 99
Holderlin, Friedrich, 147
*Holy Family* (Marx and Engel), 177–78
Holy Roman Empire, 57, 58, 59
Hugo, Victor, 7, 130, 192–93, 201
humanism, 49, 57, 59, 61–62, 68–69, 220
Humboldt, Wilhelm von, 142
Hume, David, 3, 4, 132, 140

ideology, 142, 206–8, 217, 219, 224
individualism, 4, 91, 93, 98, 100–1, 103–7, 110, 121, 123, 127, 129–30
  Kant on, 136, 141

infallibility, 95–96, 97–100, 106, 107, 109–10
Isidore of Seville, 32
Islam, 4

Jacobins, 147, 159
James (King), 101
Jansenists, 27, 122, 124, 127, 129
Jefferson, Thomas, 5
Jewish emancipation, 163–64, 171, 174, 177, 179–80
  Constituent Assembly and, 165
  German Enlightenment and, 166–67
  Hegel on, 161
  *See also* Declaration of the Rights of Man and Citizen
Joachim di Fiore, 31
John of Salisbury, 32, 34, 78
Joinville, Jean de, 60
Josephus, 32
Judaism
  assimilation and, 163, 166–67
  Christian state and, 179–80
  Dostoyevsky and, 185, 189
  and German Enlightenment, 166–67
  Marx on, 177–79
  *See also* anti-Semitism

Kant, Immanuel, 2, 3, 5–6, 24, 29, 72, 120, 122, 124, 145, 147–48, 152–53, 155–56, 162, 166, 179
  antinomies, 132, 134–35, 139, 141
  on autonomy, 138
  on "categories of understanding," 133
  on criticism, 133
  dogma, 132, 134–37
  and empiricism, 132–34, 141
  on freedom, 134, 137–38
  on individualism, 136, 141
  on knowing "thing in itself," 133, 135
  on morality, 137–39

necessity of religious practice, 132–33, 135, 139–40
  objectivity of, 134, 136
  on revelation, 139–40
  on "sensuous intuition," 133–34
  subjectivity of, 134, 136
  *See also Critique of Practical Reason* (Kant); *Critique of Pure Reason* (Kant)
Kennan, George, 217, 220
KGB, 204, 210–12, 214–17
  *See also* Soviet Union
Kissinger, Henry, 217, 218
Kjetsaa, Geir, 190
Kojève, Alexandre, 152
Khrushchev, Nikita, 210–11, 215

laity, 23–24, 26, 30–32, 37–38, 50
Le Bras, Gabriel, 86
Le Clerc, Jean, 101–2, 109
learning, of ancient world, 20
*Lectures on the Philosophy of Religion* (Hegel), 151, 153–57, 161
Leibniz, Gottfried, 126, 127, 131, 135, 137
Leninism, 203–6, 208, 217
*Lenten Letter to Pimen, Patriarch of All Russia* (Solzhenitsyn), 214–15
*Letter Concerning Toleration. See Epistola de Tolerantia* (Locke)
*Letter to D'Alembert on Theaters* (Rousseau), 118
*Letters Written from the Mountain* (Rousseau), 125
Lewis, C.S., 8, 229
*Life of Jesus* (Hegel), 148, 157
*Life of Jesus* (Strauss), 168–69
Limborch, Philip van, 101–2
Liverotto of Fermo, 67, 72
Locke, John, 2, 29, 35, 74, 84, 87, 89, 93–111, 113, 117, 122, 125, 128, 225–26, 228
  Hegel and, 151, 155, 157, 159–61
  in Holland, 101–2, 108

Locke, John—*continued*
  on infallibility, 106, 110
  Proast and, 105–6
  Roman Catholicism and, 101–5
  on Thirty Years' War, 97
  on tolerance, 97–100
  *Two Tracts on Government,* 95–97
  See also *Epistola de Tolerantia* (Locke)
*Logic* (Hegel), 149
Louis XIV (King), 94, 98, 100, 104, 120
Lowith, Karl, 150
Luther, Martin, 3, 4, 6, 73–89
  on callings, 78–79
  Charles V and, 84–85
  criticism of Church, 75–80
  excommunication, 75, 76, 79
  on faith, 73, 76, 78, 86–88
  on fragmentation of Christendom, 74
  on governing, 82–89
  on grace, 76, 77, 79–80, 86, 88
  on justification, 76, 78, 79
  nonresistance, 83–84
  on pope as Antichrist, 79, 85
  on Scripture, 80–82
  on social change, 80–81
  on worldly ranks, 78–79
Lutheranism, 96, 103, 167, 169, 178

Machiavelli, Niccolò, 2, 39, 49, 55, 219, 225
  on Christianity, 62–66
  on Church, 68–70
  criticism of, 58–60
  secularism and, 70–72
  *virtù* and, 57–58, 59–60, 63–68
Magnus, Albertus, 26
*Malebranche, 133, 135
Manichaeanism, 13–15, 17, 21, 26
Mansfield, Harvey, 58–59
Markus, R.A., 21
marriage, 74, 77–78
Marsilius of Padua, 57, 70, 100

Marx, Karl, 2, 3, 6–8, 58, 71, 145, 155, 163–80, 226–28
  anti-Semitism, 177–79
  on Augustine, 171
  Bauer and, 173–77
  on Christian state, 170–72, 174, 178–80
  on emancipation, 174–75
  Feuerbach's influence on, 169
  on Holy Alliance, 171–72
  on Jewish emanicipation, 163–67, 171, 174–75, 177, 179–80
  on species-being, 163, 169, 175–79
  See also communism
Marxism, 163, 174, 178, 180, 204, 217–18, 221, 223
materialism, 6, 123, 126, 163, 168, 172, 181, 188, 191, 197
Maury, Jean Siffrein, 165
McGrory, Mary, 221
Medici rule (Italy), 57–58, 61, 66, 68
*Meditations* (Descartes), 23
Mendelssohn, Moses, 166–67
Middle Ages, 3, 23–24, 30, 36
  secularism and, 225–26, 228
Milton, John, 41
modernity, 145, 149
*Monarchy* (Dante), 39, 44–50
  Roman Empire in, 47–48
Montaigne, Michel, 4, 87, 89, 94
Mother Teresa, 227, 229
Muentzer, Thomas, 81–82

Napoleon, 148–50, 158, 160
natural law, 32, 36–37, 57, 60, 70
  Kant on, 132, 142
Nechaev, Sergei, 188–90, 196, 198–99, 206
Nietzsche, Friedrich, 3, 7, 58, 71
nihilism, 3, 7, 181, 183–84, 187–91, 193, 195–97, 198, 200, 201
*Notebooks* (Dostoyevsky), 189–91

Oates, Titus, 99, 100
*On Christian Doctrine* (St. Augustine), 19
*On Temporal Authority* (Luther), 82–83
*On the Jewish Question* (Marx), 7, 163, 171, 173
"On the Jews and Their Lies" (Luther), 80
*One Day in the Life of Ivan Denisovich* (Solzhenitsyn), 204, 208, 214
Opus Dei, 229
Organs, 204–5, 207
original sin, 121, 123, 125
Orosius, 20, 24
orthodoxy, 3, 4, 23, 227, 228

Palmer, R.R., 3
pantheism, 7, 152, 155
papacy, 3–4, 33, 43–49, 55, 64, 66, 70
    Aquinas and, 25, 32–33
    Dante and, 46–47, 48–50
    decretals, 75
    Dostoyevsky and, 185
    empire and, 48–49
    Hegel and, 158
    Locke and, 99, 101
    Luther and, 74–75, 79, 85
    Machiavelli and, 65–70, 72
    Marx and, 171–72
    *plenitudo potestatis,* 225
    Rousseau and, 124
    Tudor repudiation of, 99
    *See also* Avignon papacy
parliamentarism, 93, 101
Pelagianism, 14–15, 21, 61
*permixtio,* 17, 19, 21, 38, 225, 229
Petrarch, 61, 63
*Phenomenology of Spirit* (Hegel), 148–49, 153–54, 159–60
Philip IV (King), 44, 46
*Philosophy of Right* (Hegel), 149, 158, 160, 161
    Marx's criticism of, 168
Pietists, 170, 174

Pizan, Christine de, 60
Plato, 24, 28–29, 36
pluralism, 4
*Pochvennichestvo,* 185
Pocock, J.G.A., 59–60
*Polycraticus* (John of Salisbury), 34
Pope Alexander IV, 25
Pope Alexander VI, 65–66, 67, 71
Pope Boniface VIII, 44, 46, 53, 60
    issuance of *Unam sanctam,* 44, 46
Pope Celestine V, 44
Pope Clement VII, 68
Pope Gregory VII, 31
Pope John XXII, 70
Pope John Paul II, 2, 228
Pope Julius II, 66
Pope Nicholas II, 30
Pope Nicholas III, 53
Popish Plot, 100–1
Popkin, Richard, 86
Popple, William, 101
*Possessed, The. See Demons* (Dostoyevsky)
postmodernism, 8
poverty, 31–32
Presbyterianism, 93, 95, 104
Prierias, Sylvester, 74
*Praise of Folly* (Erasmus), 76
*Prince, The* (Machiavelli), 58–59, 63–64, 65, 66–68, 70–71
Proast, Jonah, 105
*Profession of Faith of the Savoyard Vicar* (Rousseau), 124–26, 139
*Prolegomena to Any Future Metaphysics* (Kant), 132
Protestantism, 4–5, 27, 37, 77, 82, 86–88
    callings, 78–79, 158
    Hegel and, 151, 157–58
    Kant and, 132
    Locke and, 94–95, 96, 101, 103, 108–9
    Luther and, 77–78, 82, 86–87, 88
    Rousseau and, 122–23, 125, 128

Pufendorf, Samuel, 117, 126
Puritanism, 93–94, 95

Raab, Felix, 60
Rabb, Theodore, 100
rational theology
rationalism, 148, 157
realist-nominalist controversy, 24
reason, 23, 25–26, 29, 35–37
*Reasonableness of Christianity* (Locke), 98, 108, 109–10
Reformation, 4, 73–89, 93–95, 98, 108–10
   England and, 74–75
   Hegel on, 158–59
   Luther and, 73–74, 76, 80–82, 85–87, 88
   peasants and, 78–81, 85
   Radical, 81–82, 88
   scripture and, 74–76, 80–82, 86, 88
regicide, 34
religion
   natural, 139
   public life and, 1–2
   and the state, 1
   vocations, 25, 32
*Religion within the Boundaries of Mere Reason* (Kant), 139, 140
religiousness, 25, 28, 30–32, 35, 37–38
Renaissance, 15, 20, 26, 28, 32, 39–40, 42, 43–44, 47, 49, 51–52, 54
revelation, 23, 26–27, 28–29, 36–37, 115, 122, 124, 126–27
*Revolutionary Catechism*, 188–89
Robespierre, Maximilien, 129, 130, 139
Roman Catholic Church, 4–5, 94, 102, 165–66, 172, 173, 178
   Anti-Catholicism, 99
   Luther and, 73–74, 76–82, 84, 86–88
   Dostoyevsky and, 185–86, 189, 201–2
   Rousseau on, 113–15, 122–28

Rome
   Christianity and, 16–17, 18, 19, 20–21
   sacking of, 11, 16–17
Rousseau, Jean-Jacques, 2, 29, 33, 71, 91, 113–30, 133, 139, 175
   civic religion, 127, 129–30
   on civilization, 115–17
   on "general will," 121, 130
   illegitimate births and, 119, 121
   and nature, 118–20, 123–24, 126, 129
   on "natural religion," 119, 124, 126–27
   and sentiment, 119, 124, 126–27
   social contract theory, 117, 119, 121–22, 125, 127, 129–30
Ruge, Arnold, 171, 173
Russia, 7–8
Russian Europeanism, 194
Russian Orthodoxy, 183, 186, 190–91, 201

sacramentalists, 77, 79
*saecularis litteratura*, 16
*saeculum*, 1, 2, 11, 14–18, 20–21
Saint-Simon, Henri de, 226
Sakharov, Andrei, 217
samizdat, 211, 215
sanctity, 32, 36–37
Sartre, Jean-Paul, 204, 208, 214, 217
Savonarola, Fra, 57, 60, 62, 70, 71
Schelling, Friedrich von, 147, 150, 153, 155, 170
Schleiermacher, Friedrich, 150
Schlesinger, Arthur, Jr., 221
scientism, 181, 188
secular law, 32–33
secular politics, 57, 70, 72
secular state, 104–5, 107
self-consciousness, 6
separation of church and state, 1–2
September 11, 2001, 8

Shaftesbury, Lord Ashley, 101
Simon, Richard, 108
skepticism, 2, 3, 4, 85–87
Slavophilism, 181, 184, 201
social class, 6
*Social Contract* (Rousseau), 5, 121–22, 125
   civic religion in, 127, 129–30
socialism, 185–88, 191, 193, 195, 199, 201–2
Socinianism, 101, 108
Socrates, 156–57
*sola gratia,* 76
*sola scriptura,* 76, 98, 100
Solzhenitsyn, Aleksandr, 2, 3, 7, 181–82, 203–24, 227
   America and, 204, 213, 216–22
   on freedom, 206–7, 216, 218, 222–24
   on ideology, 206–8, 217, 219, 224
   Harvard address, 219, 221–22
   international acclaim, 212–16
   "invisible allies" and, 211, 224
   KGB and, 204, 210–12, 214–17
   Nobel Prize, 213–14
   *Novy Mir* and, 209–11, 213–1
   samizdat and, 211, 215
   on Supreme Entity, 220
   Templeton Prize, 222
   writing of *Ivan Denisovich,* 209–12
   *See also* gulags
Soviet Union, 204, 210, 213–15, 216–19
   Committee on Religious Affairs, 215
Spartans, 119
species being, 6
speculative thinking, 148–50, 152, 156
Speshnev, Nicholas, 187
Spinozism, 150
Spiritual Franciscans, 46, 48, 54, 74
St. Anselm, 23, 26, 29

St. Augustine, 2–3, 11, 13–22, 76, 83, 228–29
   combating of heresies, 14–18
   "historical" writing, 16
   on paganism, 18–22
St. Bede, 20
St. Bonaventure, 35, 53
St. Cyprian, 16
St. Jerome, 15–16
St. Louis IX, 26, 32, 33
St. Paul, 16, 18, 20
   Epistle to Romans, 73, 80
   on marriage, 78
St. Peter, 16, 18
St. Thomas Aquinas, 2–6, 23–38, 39, 60, 63, 152, 153, 156, 157, 228
   condemnation by University of Paris, 24
   Dante and, 41, 43, 47, 49, 50–51, 55
   on government, 32–35
   on happiness, 34, 35–38
   on laity, 30–32
   on religious life, 25–30
Stahl, Friedrich Julius, 170
Stalin, Joseph, 3, 7, 203–5, 207–8, 210, 215
Stirner, Max, 168, 173, 186, 193
Stoicism, 19
Storm and Stress movement, 135
Strauss, David, 168–70, 179
Strauss, Gerald, 85
Strauss, Leo, 60, 72
Stuart Restoration, 74, 93–94, 98, 101
*Summa contra Gentiles* (Aquinas), 3, 25, 28–29
*Summa theologica,* (Aquinas), 3, 27–28
Swift, Jonathan, 116

temporal power, 4
Ten Commandments, 32, 36–37
Tertullian, 15–16
terrorism, 7, 8
   *See also* September 11, 2001

Test Acts, 101, 107
Tetzel, 74
theodicy, 126, 147
Theodosius, 20
theology, 23, 26–29, 31, 35, 37–38
  Aquinas and, 28–29, 37–38
  Locke and, 102, 109
  Luther and, 79, 81, 85, 87
Thirty Years War, 58, 97, 100
Thomas, Keith, 86
Thoynard, Nicolas, 102
Tillotson, John, 108
Tindal, Matthew, 109, 127
Tocqueville, Alexis de, 182, 201
tolerance, religious, 18, 21–22, 93–95, 98–99, 100–5, 106–7, 109–10
  Judaism and, 164, 166, 180
Toleration Act, 104
totalitarianism, 181, 198, 226–27
Tracy, Destutt de, 142
Tudor Reformation, 99
Turgenev, Ivan, 183, 187, 193, 198
Tvardovsky, Aleksandr, 209–10, 213–14
*Two Tracts on Government* (Locke), 95, 100
*Two Treatises on Civil Government* (Locke), 98, 101, 103

*Unam sanctam*, 44, 46, 60
"unhappy consciousness," 158, 160
United States, 204, 213, 216–22
universities, 23–26, 28–29, 36

Varro, Marcus, 18
Vatican Council II, 229
*Vindication of the Reasonableness of Christianity* (Locke), 98, 108, 109–10
*virtù*, 57–58, 59–60, 63–68
*Vita nuova* (Dante), 42–43
Voltaire, 3, 4, 5, 113, 116, 122, 129, 133, 135, 139, 141
Von Sickingen, Franz, 80
Von Stein, Karl, 160
Von Strauss, Gerald, 85

*Warning to the West* (Solzhenitsyn), 224
Weil, Simone, 227
westernization, 181, 183, 184, 186–87, 190–92
*What Is to Be Done?* (Chernyshevsky), 183, 187–88, 194
William of Moerbeke, 24
William of Ockham, 42, 55, 70
Wolff, Christian von, 131, 135
World War II, 218, 219, 221, 222

Zwingli, 74, 77, 79, 81, 85